LONDON MATHEMATICAL SOCIETY LECTURE NOTE SERIES

Managing Editor: Professor I.M.James,
Mathematical Institute, 24-29 St Giles, Oxford

London Mathematical Society Lecture Note Series. 52

Combinatorics

Proceedings of the Eighth British Combinatorial Conference,
University College, Swansea 1981

Edited by

H.N.V. TEMPERLEY ScD
Professor of Applied Mathematics,
University College, Swansea

CAMBRIDGE UNIVERSITY PRESS

CAMBRIDGE

LONDON   NEW YORK   NEW ROCHELLE

MELBOURNE   SYDNEY

CAMBRIDGE UNIVERSITY PRESS
Cambridge, New York, Melbourne, Madrid, Cape Town, Singapore, São Paulo

Cambridge University Press
The Edinburgh Building, Cambridge CB2 8RU, UK

Published in the United States of America by Cambridge University Press, New York

www.cambridge.org
Information on this title: www.cambridge.org/9780521285148

First published 1981
Re-issued in this digitally printed version 2007

A catalogue record for this publication is available from the British Library

ISBN 978-0-521-28514-8 paperback

CONTENTS

To the memory of Dr. Derek Waller

PREFACE

This volume contains the nine invited lectures given at the Eighth
British Combinatorial Conference held in the Mathematics Building,
University College, Swansea July 20th-24th 1981. Authors of contributed
papers are making their own arrangements about publication but a booklet
of abstracts was available at the Conference and those needing further
information should write direct to the authors.

The volume is dedicated to the memory of Dr. Derek Waller, Lecturer in
Pure Mathematics at Swansea, who would have been the Conference Organiser.
In his short career he had already made significant contributions to
graph theory.

H.N.V. Temperley

ON THE ABSTRACT GROUP OF AUTOMORPHISMS

László Babai
Dept. Algebra, Eötvös University,
Budapest, Hungary.

ABSTRACT

We survey results about graphs with a prescribed abstract group of automorphisms. A graph X is said to *represent* a group G if Aut·X $\cong$ G. A class *c* of graphs is (f)-*universal* if its (finite) members represent all (finite) groups. Universality results prove independence of the group structure of Aut X and of combinatorial properties of X whereas non-universality results establish links between them. We briefly survey universality results and techniques and discuss some non-universality results in detail. Further topics include the minimum order of graphs representing a given group (upper vs. lower bounds, the same dilemma), vertex transitive and regular representation, endomorphism monoids. Attention is given to certain particular classes of graphs (subcontraction closed classes, trivalent graphs, strongly regular graphs) as well as to other combinatorial structures (Steiner triple systems, lattices). Other areas related to graph automorphisms are briefly mentioned. Numerous unsolved problems and conjectures are proposed.

0. AUTOMORPHISM GROUPS - A BRIEF SURVEY

In two of his papers in 1878, Cayley introduced what has since become familiar under the name "Cayley diagrams": a graphic representation of groups. Combined with a symmetrical embedding of the diagram on a suitable surface, this representation has turned out to be a powerful tool in the search for generators and relators for several classes of finite and finitely generated groups. This approach is extensively used in the classic book of Coxeter and Moser [CM 57] where a very accurate account of early and more recent references is also given.

Automorphism groups of combinatorial objects with high symmetry (projective spaces, block designs, more recently strongly regular graphs, two-graphs, generalized polygons) have always played an important role in group theory (cf. [Bi 71], [Ka 75], [ST 81], [Ti 74]). A study of graphs satisfying certain strong symmetry conditions was initiated, well ahead of time, by Tutte's fundamental paper on s-transitive trivalent graphs [Tu 47]. A graph is s-transitive if its

automorphism group is transitive on arcs of length s. The classification of finite simple groups, which rumour says has recently been
completed, has a substantial effect on the study of highly symmetrical
graphs. One example is the result of R. Weiss [We 81] who has shown
that the bound s ≤ 7 holds for s-transitive graphs of arbitrary valence
≥ 3 assuming the known list of 2-transitive permutation groups is
complete. (The completeness of this list follows from the classification of finite simple groups [CKS 76], cf. [Ca 81].)

Another class of graphs with a high degree of symmetry are rank-3
graphs and more generally distance-transitive graphs. These are graphs
whose automorphism group acts transitively on the set of pairs of
vertices at any fixed distance. It is this class of graphs to which,
ever since the beautiful paper of Hoffman and Singleton [HS 60], methods
of linear algebra have most successfully been applied. The method is
to derive regularity conditions (in terms of numerical parameters of
the graphs) from the symmetry conditions and subsequently to translate
the combinatorial regularity into information on the eigenvalues of
associated matrices. A detailed exposition of some of the highlights
of this theory can be found in N. Biggs' excellent book [Bi 74]. More
recent results involving such methods include the Cameron-Gol'fand
theorem which describes all 5-homogeneous graphs ([Ca 80], [Gnd 78];
cf. [Sm 75] and [CGS 78]). (A graph is k-homogeneous if any isomorphism
between induced subgraphs on at most k vertices extends to an automorphism of the graph.) Cameron has extended this result to distance-
regular graphs (under the condition of metrical 6-homogeneity) [Ca 80],
cf. [Ca x]. In fact the actual results are much stronger inasmuch as
they assume only k-regularity, a combinatorial condition which appears
to be much weaker than k-homogeneity. (For k = 2, k-regular graphs
are strongly regular; metrically 2-regular graphs are distance regular.
These conditions do not imply the presence of any non-identity automorphism.) Gol'fand has now (privately) announced the classification
of all ultra-homogeneous association schemes. (Ultrahomogeneous means
k-homogeneous for all k.)

Colored, directed graphs are a natural object on which uniprimitive
(primitive but not doubly transitive) groups act. The colors correspond
to the orbits on pairs of elements. A nice introduction to the ideas
derived from this representation is given in [Ne 77]. Besides powerful
matrix methods ([FH 64], cf. [Hi 75a,b]) which are a common generalization of the eigenvalue techniques used for association schemes [BS 52],
[De 73] and of group representation theory, there still seems to be a
lot of room left for elementary graph theoretic considerations. The
colored digràphs satisfying certain regularity conditions implied by
(but, of course, not equivalent to) primitive group action are called

primitive coherent configurations. A combination of inequalities involving valence and diameter of constituent digraphs (color classes) of such configurations have led to substantial progress on a classical problem in group theory [Ba 81]: *If G is a uniprimitive permutation group of degree* n *then* $|G| < \exp(4\sqrt{n} \log^2 n)$. (This result is sharp up to a factor of $4\log n$ in the exponent. The best previous bound was Wielandt's $|G| < 4^n$ [Wi 69].)

The results indicated in the above paragraphs have dealt with graphs satisfying regularity rather than symmetry conditions and are therefore not affected by progress in group theory. If however we assume our graphs have a high degree of symmetry (in terms of automorphisms), the classification of finite simple groups becomes relevant. Cameron [Ca 81] mentions that J. Buczak has determined all 4-homogeneous graphs, making a strong use of the list of finite simple groups. Cameron himself [Ca 81] has determined all primitive permutation groups of degree n and of order greater than $n^{(1+\varepsilon)\log n}$ ($\varepsilon \to 0$ while $n \to \infty$)(assuming the classification). It turns out that all such groups act as subgroups of S_k wr S_m on the set of ordered m-tuples of t-subsets of a k-set where $n = \binom{k}{t}^m$. One can naturally define an association scheme on this set (a combination of the Hamming and Johnson schemes). Our "large" primitive group acts on such a scheme. Such results may motivate us to ask whether these are the only association schemes with large automorphism groups. In particular, *is it true for some constant* C *that if* $|\text{Aut } X| > n^{C\log n}$ *for a strongly regular graph* X *then* X *is either complete multipartite, or the line graph of a complete or of a complete bipartite graph, or the complement of such a graph?*

Weaker conditions of symmetry such as a vertex-transitive automorphism group have interesting consequences on the structure of the graph (cf. Problems 13-19 in [Lo 79, Ch 12]). Although only four non-Hamiltonian connected vertex-transitive graphs are known, I don't feel tempted to suggest that there are only finitely many of them. It seems rather that a lack of good methods to prove non-Hamiltonicity hinders construction of an infinity of examples.

An interesting class of vertex-transitive graphs are provided by Cayley graphs. Since Maschke's paper [Ma 1896], continuing attention has been given to their embeddings on surfaces ([CM 57], [Wh 73]), but little has been done to explore other graph theoretic properties of Cayley graphs. A study of subgraphs and of the chromatic number of Cayley graphs has been initiated by [Ba 78a,b]. It is an open question whether *there is a constant* C *such that the chromatic number of any Cayley graph of a finite group with respect to an irredundant set of generators is less than* C.

Cayley diagrams play an essential role in constructing graphs with a prescribed automorphism group. The basic question, which groups are isomorphic to the automorphism group of a graph, was raised by König [Kö 36, p.5]. One can ask the same question for all kinds of mathematical structures in the place of graphs. This is what B. Jónsson [Jo 72] calls the *abstract representation problem*. Several significant directions of research on this problem find their roots in the work of R. Frucht. First, Frucht proved that *every finite group is isomorphic to the automorphism group of some finite graph* [Fr 38]. Then he went on proving that one can even require these graphs to be *trivalent* [Fr 49]. In another paper [Fr 50] he gave examples how to use graphs to construct *lattices* with given automorphism groups. Furthermore, in all these papers he was very much concerned *about the size of the graphs* he had constructed. There is one more classical paper on the subject: Birkhoff proved that every group is represented as the (abstract) group of automorphisms of a *distributive lattice* [Bi 45]. These results were followed by countless further *universality results*: constructions which prove that every group is isomorphic to the automorphism group of some object in a given class. Characteristic for the proofs is that the structure of the groups plays little role. The extent to which group structure could be ignored was forcefully demonstrated by results of Pultr, Hedrlin and their colleagues who found that most results generalize to *endomorphism semigroups* (even to categories).

Although there is a large number of papers proving independence of the abstract group Aut X and various properties of the object X (graph, ring, lattice, etc.), little has been done to establish links between properties of X and the structure of Aut X.

The aim of the present note is to survey some aspects of the abstract representation problem, with an emphasis on the (missing?) link. By presenting a large collection of open problems, we try to draw attention to the other side of a coin, one side of which has already been studied in great detail.

Basic definitions will be given in Section 1. We give a brief survey of some interesting universality results in Section 2. Contraction-closed classes of graphs are investigated in Section 3. Such classes generalize the notion of graphs embeddable on a given surface. We give a detailed proof of the fact that most classical simple groups are not represented as the (abstract) automorphism group of any graph in such a class. This is an example of the non-universality results we seek. The proof involves a study of contractions of certain Cayley graphs. The "Subcontraction conjecture" 3.3 is hoped to lead to a deeper understanding of the graph structure forced by group action. Non-universal classes of lattices are the subject of Section 4, where attempts to link combinatorial parameters such as order dimension to the automorphism group are described. Section 5 is devoted to the problem of minimizing the size of graphs representing a given group. In order to illustrate the methods available, we outline the proofs of both upper and lower bounds. Clearly, refinement of universality proofs yield upper bounds. Correspondingly, lower bounds are scarce. The "Edge-orbit Conjecture" 5.21 may indicate the possibility of an interesting lower bound. Related results and problems on lattices conclude Section 5.

Vertex-transitive representations of a given group are the subject of Section 6. The graphical and digraphical regular representation problems belong to this section. In contrast to Section 5 which is dominated by open problems, in this section we are able to survey many fine results.

Finally in Section 7 we consider the extension of some of the problems treated to endormophism monoids. We outline the proof of the basic representation theorem, keeping the minimizing problem of the number of vertices in mind. After a brief statement of some of the important universality results we turn our attention once again to what we feel ought to be an organic part of the theory but is almost entirely missing, links between the structure of the graph X and of its (abstract) endomorphism monoid.

The reader wishing to obtain a broader view of the area of graph automorphisms is referred to the excellent survey by P. Cameron [Ca x]. With that paper being in print, I have elected to write on this more compact subject which presents many problems of combinatorial rather than group theoretic nature. From among those areas closely related to graph automorphisms but omitted from this introduction let me mention the algorithmic complexity of graph isomorphism (cf. [Ba 80c]). The recent breakthrough by E.M. Luks [Lu 80] signals the relevance of the structure of the automorphism group to a depth where even the classification of finite simple groups may become relevant.

1. DEFINITIONS, NOTATION

A digraph is a pair $X = (V,E)$ where $V = V(X)$ is the set of vertices and $E(X) = E \subseteq V \times V$ is the set of directed edges. A graph is a digraph with $E = E^{-1}$ and with no loops. K_n stands for the complete graph on n vertices. A *vertex-coloured* (di)graph is a (di)graph X together with a function $f : V \to K$ which maps V into some sets of colors. (f is not a good coloring in the sense of chromatic graph theory.) An *edge-coloured* digraph is a set V together with a family of binary relations $E_1,\ldots,E_m \subseteq V \times V.(1,\ldots,m$ are the edge-colors.)

An automorphism is an X → X isomorphism. Automorphisms preserve colors by definition.

Let G be a group and H a set of generators for G. We define the *Cayley diagram* Δ(G,H) to be an edge-colored digraph with vertex set G. Colors are members of H. There is an edge of color h ∉ H joining g to gh for every g ∈ G.

The *Cayley digraph* Γ(G,H) has the same set of vertices and edges but no colors. (Therefore, it may have more automorphisms than Δ(G,H).) Γ(G,H) is a *Cayley graph* if $H = H^{-1}$ and 1 ∉ H. Cayley graphs are connected. Finite Cayley digraphs are strongly connected.

Throughout the paper, G denotes a group, n the order of G and d the minimum number of generators of G.

Both the beginning and the end of proofs are indicated by □.

2. PRESCRIBING THE ABSTRACT GROUP

In this section we consider the following type of problem: Given a group G find a graph X (or a block design, a lattice, a ring, etc.) such that the automorphism group Aut X is *isomorphic* to G. Such an object X will be said to *represent* the group G. A class *c* of objects is said to represent a class *G* of groups if, given G ∈ *G* there exists X ∈ *c* such that Aut X ≅ G. We call *c* *universal*, if every group is represented by *c*. We say that *c* is f-universal if every *finite* group occurs among the groups represented by finite members of *c*.

The first natural question was put by König in his classic monograph [Kö 36, p.5]: Which groups are represented by graphs? Frucht has soon settled the question for the finite case, proving that (the class of) graphs is f-universal. In other words:

THEOREM 2.1 [Fr 38] *Given a finite group* G *there exists a finite graph* X *such that* Aut X ≅ G.

Frucht's proof has since become standard textbook material [O 62], [Ha 69], [Lo 79], [Bo 79]. The idea is (i) to observe that the automorphism group of the colored directed Cayley diagram of G w.r. to any set of generators is isomorphic to G; (ii) to get rid of colors and orientation by replacing colored arrows by appropriate small asymmetric (automorphism free) gadgets. The same trick applies to infinite groups. We need many asymmetric graphs for that; they can be obtained from a well-ordered set by adding Frucht type gadgets between any pair a < b. We conclude that Frucht's theorem extends to the infinite case:

THEOREM 2.2 *Graphs are universal*.
(See [Bi 45], [Gr 59], [Sa 60].)

The next problem that arises is to find subclasses of graphs and classes of other (combinatorial, algebraic, topological) objects that are universal. This direction was again initiated by Frucht's fundamental discovery:

THEOREM 2.3 [Fr 49]. *Trivalent graphs are f-universal.*

Although, as Professor Frucht informs me, M. Milgram has found some gap in the original proof, there are several proofs available now (see e.g. [Lo 79, Ch. 12, Problem 8] or Section 5 of this paper). [Fr 49] had a great impact; its merit was not just proving a theorem but giving a new insight - an accomplishment offered by few flawless papers.

Another fundamental result of this kind was published a few years earlier in Spanish by G. Birkhoff:

THEOREM 2.4 [Bi 45]. *Distributive lattices are universal.*

These surprising results already foreshadow the onesidedness of later development. Take almost any interesting class of combinatorial or algebraic structures; this class is universal. (There are easy exceptions: trees, for instance. Automorphism groups of finite trees have been characterized by Pólya [Pó 37] as repeated direct and wreath products of symmetric groups. The idea goes back to Jordan [J 1869] who counted tree automorphisms. As for infinite trees, it is easy to see that if a finite group is represented by an infinite tree then it is represented by a finite one as well. - Groups are not universal either: it is easy to see that for no group G ($|G| \geq 3$) is Aut G a cyclic group of odd order.)

Some of the universality results are straightforward. Bipartite graphs are an example. Just take a connected graph X which is not a cycle, and halve each edge by inserting a new vertex of degree two. The obtained graph Y is bipartite and satisfies Aut Y $\cong$ Aut X. Now, use Theorem 2.2 to prove that bipartite graphs are universal.

It is quite easy to prove that Hamiltonian, k-connected or k-chromatic graphs are f-universal; it is somewhat more difficult to extend Frucht's Theorem 2.3 to regular k-valent graphs (k $\geq$ 3). Sabidussi's paper [Sa 57] proving that these and some other classes of graphs were f-universal, was soon considered as compelling evidence to support the view that *"requiring X to have a given abstract group of automorphisms was not a severe restriction"* [Ha 69, p. 170], [O 62, Ch 15.3]. Izbicki proved that certain combinations of Sabidussi's conditions are still insufficient to restrict the auto-morphism group [Iz 57, 60]. Universality results in algebra and topology were inspired by de Groot's papers [Gr 58, 59]; one of his results there is that commutative rings are universal. A surprisingly strong version of this was given by Fried and his undergraduaate student Kollár:

THEOREM 2.5. [FK79,81] *Fields are universal*

Finite extensions of $\mathbb{Q}$ are universal over finite groups. (We have to note here that these extensions are not normal. Noether's classical question whether every finite group is the Galois group of a polynomial over $\mathbb{Q}$ remains open, see [Sh 54].)

From the numerous universality results for finite combinatorial structures, let me quote some appealing ones. They are due to E.Mendelsohn:

THEOREM 2.6 [Me 78a]. *Steiner triple systems as well as Steiner quadruple systems are f-universal.*

COROLLARY 2.7 [Me 78b]. *Strongly regular graphs are f-universal.*

In order to see how 2.7 follows from 2.6, let X be a Steiner triple system. Take its line graph L(X) (vertices are triples from X, adjacency means non-empty intersection). Points of X correspond to maximum cliques in L(X). Conversely, every maximum clique of more than 7 vertices in L(X) corresponds to a point in X. (This follows, for instance, from Deza's theorem [De 74] (see [Lo 79, Ch. 13 Probl. 17]): if the intersection of every pair of more than n^2-n+1 n-sets are of the same size, then these pairwise intersections coincide.) We conclude that for $|V(X)| > 15$, one can recover X from L(X) hence Aut X $\tilde{=}$ Aut L(X) proving 2.7.

The proofs of 2.5, 2.6 and many other similar results start from a graph X with given automorphism group, and build an appropriate object X' such that Aut X $\tilde{=}$ Aut X'. It is usually easy to find X' such that Aut X $\tilde{\leq}$ Aut X'. The task is then to make it sure that X' has no superfluous automorphisms. There are interesting cases, however, where even the "subgroup problem" is open.

PROBLEM 2.7. *Prove for every* k $\geq$ 3, *that, given a finite group* G, *there is a* BIBD *of block size* k (a 2-(v,k,1)-design) X *such that* G $\tilde{\leq}$ Aut X.

Such X is easily found if $k = p^{\alpha}$ or $p^{\alpha}+1$ (p prime): take affine or projective spaces of high dimension over $GF(p^{\alpha})$ with the lines as blocks. For such k one can in fact prove that BIBD's with block size k are f-universal [Ba y], extending the STS result of E. Mendelsohn.

CONJECTURE 2.8. BIBD's *of block size* k *are f-universal for any fixed* k $\geq$ 3.

We note that an affirmative answer to 2.7 is known when k is a multiple of the order of G [Wil].

Although there are many more interesting universality results and some open problems of this kind, the literature on them has grown out of proportion without the healthy balance of theorems that would provide links between the group structure of Aut X and the combinatorial nature of X. This situation may derive largely from the nature of the subject but to some extent also from the pressure to publish

or perish. Techniques for constructing objects with a given automorphism group are well developed, and such (sometimes easy, sometimes ingenious, often tedious) constructions dominate the subject.
There is a chronic lack of questions pointing to possible links rather
than to independence of the structure of Aut X and properties of X.
My main objective in the next two sections is to show that such links
do exist and exploring them could be a worthwhile task.

3. NON-UNIVERSAL CLASSES OF GRAPHS

Tournaments are not universal. Their automorphism groups have
odd order for the simple reason that any involution would reverse an
edge. On the other hand, every finite group of odd order can be
represented by a tournament [Mo 64] (cf. [Lo 79, Ch. 12, Problem 7]).
So this again is a universality type result rather than the kind we
seek.

Turán asked in 1969 whether planar graphs were f-universal. The
negative answer [Ba 72] was the starting point of the author's research
in this direction. Graphs embeddable on a given compact surface were
shown to be non-universal [Ba 73] and the automorphism groups of planar
graphs have been fully described in terms of repeated application of a
generalization of wreath product, starting from symmetric, cyclic,
dihedral groups and the symmetry groups of Platonic solids (A_4, S_4, A_5)
[Ba 75]. I expect that in some sense, such structure theorems should
hold under much more general circumstances. As a first step, we find
the following non-universality result.

Contraction of a graph X onto a graph Y is a map $f : V(X) \to V(Y)$
such that (i) u,v ∈ V(Y) are adjacent iff u = f(x), v = f(y) for some
adjacent pair x, y of vertices of X; (ii) the subgraph of X induced
by $f^{-1}(u)$ is connected for every vertex u of Y.

Y is a *subcontraction* of X if Y is a subgraph of a contraction of
X. Clearly, the class of finite graphs embeddable on a given surface
is *subcontraction closed*. It is also easy to see that there are subcontraction closed classes of finite graphs, not embeddable on any
compact surface. (Take the graphs with no block of more than 5
vertices, for instance.) The interest in subcontraction closed classes
stems among other things from Hadwiger's conjecture. For the graph
theory of subcontraction we refer to [Ma 68], [Ma 72], [O 67].

The principal result intended to illustrate our point is this.
THEOREM 3.1. [Ba 74a]. *If a subcontraction closed class of graphs is
f-universal then it contains all finite graphs.*

The infinite version of this is still open:
CONJECTURE 3.2. *If a subcontraction closed class of graphs is universal, then it contains all graphs.*

An equivalent formulation of this conjecture is the following: *Given a cardinal κ prove that there is a group G such that for any graph X, if Aut $\tilde{=}$ G then X contains a subdivision of the complete graph K_κ.*

A good candidate for such G might be a large alternating group or some other simple torsion group.

Non-universality results immediately call for an investigation of the structure of those groups actually represented by the class of objects in question. As a first step toward a structure theory of these groups, one might like to find out which *simple* groups are represented. My favourite problem pertains to this question.

SUBCONTRACTION CONJECTURE [Ba 75] 3.3. *Let C be a subcontraction closed class of graphs, not containing all finite graphs. Then the set of non-cyclic finite simple groups represented by C is finite.*

Another way of stating this problem is this:

CONJECTURE 3.3'. *Given an integer k find an N = N(k) such that if G is a finite simple group of composite order greater than N(k) and X is a graph such that Aut X $\tilde{=}$ G then X has a subcontraction to the complete graph K_k.*

There are partial results in this direction. In [Ba 74a,b] it is shown that the conclusion of 3.3' holds if G contains $Z_p \times Z_p \times Z_p$ (the elementary abelian group of order p^3) for some large prime. (We assume G is simple.) Another large class of finite simple groups G is taken care of by the following result. Let p and r be prime numbers, $p \equiv 1 \mod r$ and let $H(p,r)$ denote the nonabelian group of order pr.

THEOREM 3.4. *For every k there is an M = M(k) such that if the finite simple group G contains $H(p,r)$, $r > M(k)$ and Aut X $\tilde{=}$ G then X has a subcontraction onto K_k.*

□We sketch the proof which goes along the lines of [Ba 74a]. We continue the present discussion after Remark 3.11 so the reader wishing to omit proofs may turn to that page. First one proves the following lemma.

LEMMA 3.5. *If Aut X $\tilde{=}$ G is a finite simple group then X has a subcontraction Y such that (i) Y is connected, (ii) G acts as a subgroup of Y, (iii) this action is transitive on edges and (iv) no vertex of Y is fixed under the action of G.*

(A minimal subcontraction of X satisfying (i), (ii) and (iv) will also satisfy (iii).)

If Aut Y is transitive then Y is regular. Otherwise Aut Y has two orbits, Y is bipartite and semiregular (vertices in each color class have equal valences). If all vertices of Y have large valence, a theorem of Mader yields the result:

THEOREM 3.6 [Ma 68]. *If Y is a finite graph with* $|E(Y)| > g(k)|V(Y)|$
then Y has a subcontraction to K_k. *Here* $g(k) = 8\lceil k \log_2 k \rceil$ *and* $\lceil x \rceil$
denotes the least integer $\geq x$.

Another tool we shall use is the
CONTRACTION LEMMA [Ba 73], [Ba 77a] 3.7. *If a group H acts semi-*
regularly on a connected graph X then X has a contraction onto a
Cayley graph of H.

Semiregular action means that no vertex of X is fixed by any non-
identity element of G. The proof of this lemma is simple: one has to
find a maximum tree T containing at most one vertex from each orbit.
This tree will meet every orbit. By contracting T and its images
under G to one point each we obtain a Cayley graph of G.

We shall use the Contraction Lemma with $H(p,r)$ in the role of
H. To this end, we need
LEMMA 3.8. *Any Cayley graph W of* $H(p,r)$ *has a contraction to* K_k *if*
$r > 2g(k)$ *where* $g(k)$ *is as in Theorem 3.6.*

In fact we shall prove that W has a contraction onto a regular
graph of valence not less than r. By 3.6 this implies 3.8.

□ Let W be the Cayley graph corresponding to a set S of generators
of $H(p,r)$. The set S contains an element h of order r and an element
g not in the cyclic subgroup C generated by h. Now W contains the
edges {u,uh} u∈H(p,r), therefore the left cosets uC induce connected
subgraphs. Let us contract them to one point each. The resulting
graph on p vertices admits a cyclic automorphism of order p and is
therefore regular. The number of neighbors of the set C is at least
r since the vertices $h^i g$ (i = 0,...,r-1) belong to different left
cosets of C. This proves 3.8. □

Lemma 3.8 and the Contraction Lemma together with Mader's theorem
tell us that we are done if $H(p,r)$ acts semiregularly on Y. In the
opposite case, however, Y necessarily contains vertices of valence not
less than r. For the following is straightforward.
PROPOSITION 3.9. *Let s be a prime number and Y a connected graph. If*
an automorphism of order s fixes a vertex of Y then it also fixes one
of valence not less than s.

Hence we are done if Aut Y is transitive. In the bipartite case
we still need one more trick.
PROPOSITION 3.10. *Let Y be a finite semiregular bipartite graph. Let*
A and B be the two color classes with each vertex in B having valence
b. Let Z denote the graph with vertex set A where two vertices are
adjacent iff they have a common neighbor in B. Then Y has a contraction
onto a graph $Z' = (A,E)$ *such that* $|E| \geq |E(Z)|/\binom{b}{2}$.

□ For let $B = \{v_1,\ldots,v_m\}$ and let us partition $E(Z)$ as $E_1 \cup \ldots \cup E_s$
where E_i consists of those pairs $u_1,u_2 \in A$ having v_i but none of

$v_1, \ldots, v_{i-1}$ for a common neighbor. Clearly $|E_i| \leq \binom{b}{2}$ and so an appropriate contraction f can be defined setting $f(u) = u$ ($u \in A$), $f(v_i)$ a vertex of an arbitrary member of E_i if E_i is not empty and $f(v_i)$ any neighbor of v_i of E_i is empty. The number of edges in $f(Y)$ is at least the number of non-empty sets E_i which in turn is not less than $|E(Z)|/\binom{b}{2}$, proving 3.10. □

Now we are able to finish the proof of 3.4. In the remaining cases Y is bipartite with color classes A and B. The vertices in A have valence r or more (by 3.9). The valence of the vertices in B is $b \leq 2g(k)$ (g(k) as in 3.6)(or else Mader's theorem 3.6 concludes the proof). $H(p,r)$ must not act semiregularly on A (again by 3.9). Define Z as in 3.10. Then G acts transitively on the connected graph Z and the subgroup $H(p,r)$ does not act semiregularly. It follows (by 3.9) that $|E(Z)| \geq r|A|/2$. By 3.10, Y has a contraction onto a graph $Z' = (A,E)$ with $|E| \geq |E(Z)|/\binom{b}{2} > r|A|/b^2 \geq r|A|/4g(k)^2$. Finally, an application of Mader's theorem to Z' concludes the proof of Theorem 3.4 assuming $r > M(k) \geq 4g(k)^3$. □

REMARK 3.11. We see that $M(k) = ck^3\log_2^3 k$ is sufficient for Theorem 3.4 to hold, with c some constant (less than 2000). The same proof works for $Z_p \times Z_p \times Z_p$ in place of $H(p,r)$, with the same bound $p > M(k)$. The only difference is that 3.8 has to be replaced by the easier fact that any Cayley graph of $Z_p \times Z_p \times Z_p$ is contractible onto K_p (see [Ba 74b]).

Let me add some more comments on Conjecture 3.3'. With the classification of finite simple groups in sight, a case-by-case treatment of possible simple automorphism groups does not seem inappropriate. Theorem 3.4 handles large classes of simple groups of Lie type. (Happily, we don't have to worry about sporadic groups in such asymptotic problems.) The next attack ought to be concentrated on classical groups, in particular on $PSL(n,q)$, $n \geq 2$, $q = p^\alpha$ a prime power. If either n is large, or p is a large prime and $\alpha(n+1) \geq 3$ then $PSL(n,q)$ contains $Z_p \times Z_p \times Z_p$. If p-1 has a large prime divisor r then $PSL(n,q)$ contains $H(p,r)$. Both cases are solved by 3.4 and the result quoted before 3.4. For the remaining cases we may assume n=2 and q is large. There are two separate open cases here:

    (i) $\alpha=1$ (q is a large prime) but all primes dividing q-1 are bounded.

    (ii)   p is bounded, α large.

I guess case (i) will be relatively easy.

Once the Subcontraction Conjecture 3.3 is settled, one might proceed to the following

CONJECTURE 3.12. *Let C be a subcontraction closed class of finite graphs not containing all finite graphs. Then every composition factor of* Aut X *(X∈C) is cyclic, alternating, or a member of a finite family of simple groups. (This family depends on C, of course.)*

It would be definitely premature to formulate conjectures beyond this point. In order to illustrate the kind of structure I have in mind, let me nevertheless state one more problem which has nothing to do with simple groups.

CONJECTURE 3.13. *Let C be as in 3.12. Then there is a number N such that if the order of the group G is not divisible by any primes less than N and G $\tilde{=}$ Aut X for some X∈C, then G is a result of repeated application of direct and wreath products, starting from cyclic groups.*

I believe p-groups are not easier to represent than others.

CONJECTURE 3.14. *Let C be as in 3.12. Then for every prime p there is a p-group not represented by C.*

It is essential here that p be a small prime. For large primes, it is not difficult to prove 3.14.

In the infinite case, even abelian groups may be quite complicated.

CONJECTURE 3.15. *Given a cardinal κ there is an abelian group G such that if* Aut X $\tilde{=}$ G *for some graph X then X has a subcontraction to the complete graph Kκ.*

The *Hadwiger number* h(X) of a graph X is the maximum k such that X has a subcontraction to K_k. For planar graphs h(X) ≤ 4. It would be interesting to classify finite groups G by the quantity m(G) m(G) = min{h(X) : Aut X $\tilde{=}$ G}. For finite abelian groups, m(G) ≤ 3. What is the asymptotic order of magnitude of m(PSL(2,p)) and of m(A_n)? For the latter, 3.11 guarantees that m(A_n) > c'$n^{1/3}$/log n but this may be a very poor estimate. I don't know any reasonable upper bound. Does log m(A_n) = o(n log n) hold?

After so many problems on subcontraction closed classes, let me emphasize again that the main problem is to find properly chosen classes of graphs where deep relationship between graph and group structure could be expected. Subcontraction closed classes may serve as a model case; hopefully other such classes will be discovered before long. The next section is intended to provide further hints in this direction.

4. NON-UNIVERSAL CLASSES OF LATTICES

Our aim is to find such classes.

One possible approach is to restrict our attention to a variety (equational class) of lattices (see [Gr 68], [Gr 78]). By Birkhoff's result (Theorem 2.4) however, every non-trivial variety of lattices is universal and the finite members of such a variety are f-universal.

One might also think of restricting certain combinatorial parameters of a (finite) lattice. I would like to draw the reader's
attention to two parameters. Both are sometimes called dimension.
In order to avoid confusion, we shall call one of them the *height* of
the lattice (the maximum chain length minus one); the other is the
order dimension or Dishnik-Miller dimension, defined for a poset P
as the minimum number of linear orders on the same underlying set
whose intersection is P.

The usual phenomenon occurs again. *Lattices of height 3 and
order dimension 4 are universal* [BD 81]. The simple proof goes back
to Frucht [Fr 50].

Intriguing problems arise, however, when we combine algebraic
and combinatorial restrictions. Modular lattices are universal (since
distributivity implies modularity). In contrast, here are two
conjectures.

CONJECTURE 4.1. *Finite modular lattices of bounded height are not f-
universal.*

This problem is open even in the particular case when the height
bound is 3. Modular lattices of height 3 are the subspace lattices of
projective planes. There are very strong restrictions on the automorphism groups of planes of order 4k-1 and 8k+5 [He 67], [He 72]
(cf. [De 68, Ch. 4]). The situation is more complicated for planes
of order 8k+1 [He 72].

We mention that the infinite version of 4.1 is false; (infinite)
projective planes are universal [Me 72]. The proof goes via free
completion [Ha 43] of a partial plane constructed from a graph with
prescribed automorphism group.

4.1 will also fail if modular lattices are replaced by geometric
(matroid) lattices. *Matroids of rank 3 are f-universal* [Ba 78d].

CONJECTURE 4.2. *Modular lattices of bounded order dimension are not
f-universal.*

This problem has been attacked in [BD 81] with the result that
the following additional condition is sufficient for non-universality:
intervals of height 2 have a bounded number of elements. In fact, in
such a case our lattices do not represent the cyclic group Z_p where p
is a prime number exceeding the bound on the size of height 2 intervals
and also greater than N(k) where k is the bound on the dimension and
N(k) is a doubly exponential function of k. For such p, the order of
the automorphism group is not even divisible by p, indicating how far
we are from universality - and from any conclusive result.

The nature of this problem seems combinatorial rather than algebraic.
As an illustration, let me quote one of the key propositions from [BD 81].

A hypergraph (V,E) is said to be *almost non-intersecting*, if

|E∩E'| ≤ 1 for any two distinct members E, E' of *E*. An *arc of length* k in (V,*E*) is a k-subset S of V such that |S∩E| ≤ 2 for any E ∈ *E*.
PROPOSITION 4.3. *Let* p *be a prime number and let* (V,*E*) *be an almost non-intersecting hypergraph on* p *vertices with a vertex-transitive automorphism group. If* V∤E *then there is an arc in* (V,*E*) *of length greater than* $(2p)^{1/4}$. (See [BD 81, Prop. 416].)

What happens for composite p? Is there a similar lower estimate for the maximum arc length? How far is this estimate from the correct order of magnitude? (Note that our hypergraph is not necessarily uniform.)

5. SMALL GRAPHS (LATTICES, ETC.) WITH GIVEN GROUP

Frucht [Fr 38, 49, 50] sought not only to construct graphs (posets, lattices) representing a given group, but to *minimize* their size. Little has since been done in this regard, although quite intriguing problems can be raised.

The dilemma of *constructions* (universality) vs. *theorems* (non-universality) discussed in Sections 3 and 4 correspond to finding upper vs. lower bounds on the minimum order of graphs (and other structures) representing a given group. Correspondingly, there is a discomforting *lack of lower bounds* on the minimum order. Let me start with an example. For a group G, let v(G) denote the *minimum number of vertices* of a graph representing G. What is the order of magnitude of $v(A_m)$ where A_m denotes the alternating group of degree m? It is easy to see that $v(A_m) \leq m!$ (see Theorem 5.4 below), and $v(A_m) \leq m!/2$ follows from [Wa 74]. In reality, I am sure $v(A_m) = o(m!)$. On the other hand, m! may be close to the correct order of magnitude.

PROBLEM 5.1. *Is* $v(A_m) > m^{cm}$ *for some positive constant* c?

I am not quite sure about this, but $v(A_m) > 2^{cm}$ seems certain although it has not yet been proved. Bochert's estimate on the order of primitive permutation groups [Wi 64, p. 41] will be helpful in attacking this problem. It rules out the cases when the stabilizer of a vertex is a primitive group if regarded as a subgroup of A_m in its natural representation of degree m.

Some particular classes of groups (cyclic, dihedral, symmetric groups, direct products and certain wreath products of these groups) have been considered in detail (see the survey [MQ 79].) In the following discussion we are interested in more complex classes of groups.

For a group G let orb(G) *denote the minimum number of orbits* of Aut X for those graphs X satisfying Aut X ≅ G. Let further d = d(G) denote the minimum number of generators of G. The order of G will henceforth be denoted by n. Clearly,

$$d \le \log_2 n$$

Another straightforward observation is the following:

PROPOSITION 5.2. $v(G) \le n \; orb(G)$. (Each orbit has length $\le n$.)

With the exception of the particular classes of groups mentioned the known bounds for $v(G)$ are proved by constructing a graph X representing G such that G acts *semiregularly*. In such a case each orbit has length n hence $|V(X)|$ equals n times the number of orbits of Aut X. Therefore these bounds can actually be stated in a slightly stronger form, as bounds on orb(G).

[Fr 38] gives an $O(d^2)$ bound on orb(G). This was improved in [Fr 49] to $O(d)$, subsequently by Sabidussi [Sa 59] to $O(\log d)$, and later on in [Ba 74] to $O(1)$ by means of a simple encoding of the colors and orientations of the Cayley diagram.

THEOREM 5.3 [Ba 74]. $orb(G) \le 2$ *for all finite groups except for the cyclic groups of orders* $n=3,4,5$.

We postpone the proof until the line-minimum version of this result (5.19).

This result is sharp for generalized dicyclic groups (see before 6.13) and for abelian groups of exponent greater than 2, because these groups cannot be represented by vertex-transitive graphs [No 68], [Wa 71] (cf. end of Section 6).

COROLLARY 5.4. $v(G) \le 2n$ *unless* G *is cyclic of order* $n=3,4,5$.

This is about as far as one can get entirely ignoring the structure of G. The only infinite classes of groups for which this estimate is known to be sharp are cyclic groups of prime order and generalized quaternion groups.

PROBLEM 5.5. *Are there infinitely many other groups for which equality holds in* 5.3?

The basic problem, however, is that the bound 5.4 seems very poor for most groups, the true value of $v(G)$ being substantially less than n.

It seems worthwhile to seek broad classes G of finite groups such that

(i) $\lim_{G \in \mathcal{G}} v(G)/|G| = 0$

Let c(G) denote the *maximum order of cyclic p-subgroups of* G. Clearly, G has no faithful permutation representation of degree less than c(G). Therefore the following condition is necessary for (i):

(ii) $\lim_{G \in \mathcal{G}} c(G)/|G| = 0$.

PROBLEM 5.6. *Is* (ii) *a sufficient condition for* (i)?

In fact, it is not clear whether groups satisfying (ii) have faithful permutation representations of degree o(n) at all.

An interesting subproblem of 5.6 is when G consists of p-groups for some fixed prime p. (For p-groups, c(G) is the exponent of G.) A sufficient condition for (ii) is that
(iii) the number of primes dividing G tends to infinity for $G \epsilon G$.

Another sufficient condition for (ii) is that
(iv) the minimum number of generators tends to infinity for $G \epsilon G$.

PROBLEM 5.7. *Does* (i) *follow from either* (iii) *or* (iv)?

(iv) remains interesting when G consists of p-groups for a fixed prime p. (iii) trivially implies (i) for nilpotent groups (direct products of p-groups) but for solvable groups, (iii) remains interesting.

These problems are, in essence, problems on *2-closed representations* of finite groups. A permutation group acting on a set V is 2-closed [Wi 69] if it coincides (as a permutation group) with the automorphism group of an edge-colored digraph on the vertex set V.

For any class of groups, 5.6 is equivalent to its *edge-colored digraph* version with a bounded number of colors. Let $X = (V; E_1, \ldots, E_k)$ be a k-edge-colored digraph with *disjoint colors*, i.e. assume $E_i \cap E_j = \emptyset$ for $i \neq j$. (The E_i are subsets of V×V.) Let $v_k(G)$ denote the minimum number of vertices of such a colored digraph representing G. Let $v_\infty(G) = \min v_k(G)$. Clearly $v_\infty(G) \leq v_{k+1}(G) \leq v_k(G) \leq v(G)$ for any $k \geq 1$. An adaptation of an idea of Sabidussi [Sa 59] yields the following inequality:

PROPOSITION 5.8. $v(G) \leq (v_k(G)+1)(1+2^{\lceil \sqrt{\log_2(k+1)} \rceil})$.

We shall outline the proof in the more general context of endomorphism monoids (7.3).

COROLLARY 5.9. $v(G) \leq (v_\infty(G)+1)(1+2^{\lceil \sqrt{\log_2 n} \rceil})$ *where* n = |G|.

We don't know whether this inequality can be improved on by an unbounded factor. On the other hand, we are unable to disprove the relation $v(G) = O(v_\infty(G))$.

PROBLEM 5.10. *Does there exist a constant C such that*
$$v(G) < Cv_\infty(G)$$
for all finite groups G?

Problems 5.6, 5.7 and 5.10 will require deeper insight in the structure of 2-closed permutation groups. Nevertheless, there are still interesting open problems in this area which may be amenable to a more elementary approach. Some of these concern subclasses of graphs.

First we give an example where a *satisfactory lower bound* has been found in order to complement the available upper bound. Let $v^k(G)$ denote the minimum number of vertices of a k-valent graph representing G.

THEOREM 5.11. [Fr 49] $v^3(G) = O(nd)$. (n = |G|, d is the minimum number of generators.)

We postpone the proof until after 5.17. The lower bound, too, is of the form O(nd).

THEOREM 5.12 [Ba 77a]. *Let* G *be indecomposable with respect to direct product and assume that* g.c.d.$(n,6) = 1$. *Then* $v^3(G) \geq 2n(d-1)$.

In 5.12, direct indecomposability is clearly necessary. (If G $\tilde{=}$ H×K, then the disjoint union of nonisomorphic graphs representing H and K, resp. will be smaller than the bound we seek. If we are interested in connected graphs only, the condition of direct indecomposability can be dropped.) Some kind of condition preventing G from having too many elements of order 2^α is also necessary, as shown by the connected trivalent graph X_ℓ obtained from a cycle of length 2ℓ by replacing every other edge by K_4^-, the complete 4-graph minus an edge. (See Figure 1).

Figure 1. The graph X_3

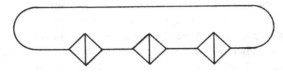

The group G = Aut X_ℓ has order n = $\ell 2^{\ell+1}$. It is generated by d $\leq$ 3 permutations. X_ℓ has 4ℓ vertices, which is O(log n).

The condition that 3 does not divide n is less important. In fact. we prove a O(nd) lower bound without this condition.

THEOREM 5.13. *Let* G *be a group of odd order, indecomposable w.r. to direct product. Then* $v^3(G) \geq 2n(d-1)/3$.

The proof of 5.12 and 5.13 rests on the Contraction Lemma 3.7, supplemented with a valency estimate.

CONTRACTION LEMMA (complete form) 5.14 [Ba 77a]. *Let* X *be a connected regular graph of valence* r. *Suppose that a group* G *acts semiregularly, with* t *orbits, on the vertex set of* X. *Then* X *is contractible onto a Cayley graph of* X *of valence* ρ,

ρ $\leq$ t(r-2) + 2.

(Let us remark that the Nielsen-Schreier theorem on subgroups of free groups and the Schreier index formula are immediate corollaries of the Contraction Lemma, cf. [Ba 77a].)

☐ Now we prove 5.13. Let X be a trivalent graph of minimum order v, representing G. Using that G is indecomposable and has odd order, we infer that X is connected. The stabilizer of an edge in the automorphism group of a connected trivalent graph is a 2-group. (This is a trivial fact; it is more difficult to prove the converse, namely that every finite 2-group can be represented as such an edge-stabilizer [BL 73].) By our condition that n is odd it follows that G acts semiregularly on the set of edges. Therefore by the Contraction Lemma the line graph L(X) has a contraction onto some Cayley graph of G of valence ρ, say. Clearly $\rho \geq 2d$. On the other hand, $\rho \leq t(r-2)+2$ by 5.14 where t is the number of edge-orbits and r=4 is the valence of the line-graph. We conclude that d $\leq$ t+1. The number of edges is tn on the one hand, 3v/2 on the other hand, hence v = 2tn/3 $\geq$ 2n(d-1)/3, proving 5.13. The proof of 5.12 goes similarly with the difference that in that case G acts semiregularly on X itself and therefore there is no need to turn to L(X). □

We expect that the following more general lower bound holds.

CONJECTURE 5.15. *To each positive integer* α *there is a positive number* $c(\alpha)$ *such that if G is a group, indecomposable w.r. to direct product and* 2^{α} *does not divide* $n = |G|$ *then* $v^3(G) > c(\alpha)nd$.

5.12 can easily be generalized to k-valent graphs. The result is this:

PROPOSITION 5.16. *Let G be indecomposable w.r. to direct product and assume g.c.d.* $(n,k!) = 1$. *Then* $v^k(G) \geq 2n(d-1)/k-2)$. *Under the weaker assumption that g.c.d* $(n,(k-1)!) = 1$ *we have* $v^k(G) \geq 2n(d-1)/k(k-2)$.

As for upper bounds, [Sa 57] shows $v^k(G) = O(knd)$. By way of some tedious but routine construction one can relatively easily improve this to $v^k(G) = O(nd)$ (k $\leq$ n)(using 5.11).

I believe however, that 5.16 is nearly sharp - at least if d exceeds k.

CONJECTURE 5.17. *There is an absolute constant* c *such that* $v^k(G) \leq cn^{\lceil d/k \rceil}$ *whenever* k $\leq$ n.

□ For completeness, we outline the *proof of the upper bound* $v^3(G) = O(nd)$ for trivalent graphs (Theorem 5.11).

Let d' denote the minimum size of a symmatric set S of generators of G (S = S^{-1}). (This is the minimum of the valences of Cayley graphs of G.) Clearly, d $\leq$ d' $\leq$ 2d. We assume d' $\geq$ 3. (Otherwise G is cyclic or dihedral. These cases are easily settled directly.) First we construct an edge-coloured mixed graph Y on d'n vertices such that Aut Y $\tilde{=}$ G and the total valence of each vertex of Y is 3. By a *mixed* graph we mean a graph which is allowed to have both directed and undirected edges. The *total valence* of a vertex x in a mixed graph is the sum of the valence of the undirected part of Y at x plus the number of directed edges entering and of those leaving x. This graph Y will be obtained by "blowing up" each vertex of the Cayley diagram $\Delta(G,S)$ to a cycle. Let S = $\{g_1,...,g_d,\}$. We set V(Y) = G×S. The vertices $(g,g_1),(g,g_2),...,(g,g_d,)$ will induce (in this cyclic order) a directed cycle (colored brown, say) for each g ϵ G. Let S be the disjoint union of the sets I, H and H^{-1} where I consists of the involutions (elements of order two) in S, and H is a set of representatives for the pairs

$\{h,h^{-1}\}$ ($h\epsilon S\backslash I$). Let us draw an undirected edge of color $h\epsilon I$ between (g,h) and (gh,h). Furthermore, let us introduce a directed edge of color $h\epsilon H$ from (g,h) to (gh,h^{-1}).

Clearly, the left translations $(g,h) \to (xg,h)$ ($x\epsilon G$) induce a group of automorphisms of Y and it is easy to see that Y has no other (color- and orientation preserving) automorphisms.

The next step is to merge most of the colors in order to reduce their number. So far, all the d' edges entering or leaving a brown cycle have pairwise different colors or orientations relative to the cycle. Let us retain just four colors.

Let us symmetrize all edges of the brown cycles (make them undirected) with the exception of the edges $((g,g_1), (g,g_2))$. Let us retain brown for these directed edges (one from each cycle), and assign black to the $n(d'-1)$ undirected edges obtained. For the rest of the edge set, we color undirected edges blue and directed edges green. From this less colorful picture one can easily recover the original colors and thereby prove that no new automorphisms arise.

As a final step we select two appropriate Frucht type directed "gadgets" and an undirected one, to replace brown, green and blue edges, resp. (See Fig. 2.)

Figure 2. Two directed and an undirected gadgets for trivalent graphs.

Starting from the triangles of the graph X obtained, one can uniquely reconstruct Y from X. Consequently, the automorphisms of X are exactly those induced by automorphisms of Y, hence Aut X $\cong$ Aut Y $\cong$ G. The number of vertices of X is $|V(X)| = n(5d'-|I|+8) < 10n(d+1)$. This proves Theorem 5.11. $\square$

Although this upper bound on $v^3(G)$ is of the same order of magnitude as the lower bound 5.12, the gap may still deserve attention. PROBLEM 5.18. Prove (or disprove): *for every positive ϵ there exists* $d(\epsilon)$ *such that* $v^3(G) < (2+\epsilon)nd$ *whenever* $d > d(\epsilon)$.

In fact, $v^3(G) < (1+\epsilon)nd'$ is the likely bound (where d' is, as in the proof above, the minimum order of symmetric sets S of generators of G ($S = S^{-1}$)).

Now we turn to the problem of *minimizing the number of edges*. Let e(G) denote the minimum number of edges of graphs representing the group G. We have an O(nd) upper bound on e(G). In fact, $e(G) \le 3v^3(G)$

hence an O(nd) bound follows immediately from 5.11. We prove now
that an O(nd) bound on the number of edges and an O(n) bound on the
number of vertices can be attained simultaneously.

THEOREM 5.19. *With three exceptions, every finite group G is
represented by a graph X such that* $|V(X)| \leq 2n+1$ *and* $|E(X)| \leq 2n(d+1)$.
(Here, as throughout, n = |G| and d denotes the minimum number of
generators of G.) The exceptions are the cyclic groups of orders 3,
4, 5.

    □ The proof consists of a slight modification of the proof of 5.3
[Ba 74c]. We may assume $n \geq 6$. For cyclic groups, one can use a
direct construction [Sa 59] (See Fig. 3).

    Figure 3. A graph for C_n, $n \geq 6$.

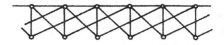

Otherwise $d \geq 2$. Let $h_1,\ldots,h_d$ be a minimum set of generators. Let
$V(X) = (G\times\{1,2\})\cup\{\omega\}$ where ω is a special vertex. Let us introduce
the following edges (for all $g\in G$): on the "first floor"
$[(g,1), (gh_1,1)]$; on the "second floor"
$[(gh_i,2),(gh_{i+1},2)]$ $(i=1,\ldots,d-1)$; between the two floors $[(g,1),(g,2)]$
and $[(g,1),(gh_i,2]$ $(i=1,\ldots,d)$. Finally, we connect ω (the "roof) to
each vertex on the second floor. (See Fig. 4.)

    Figure 4.

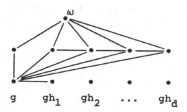

    Clearly, G acts on the obtained graph X by left translations on
both floors. It is not hard to verify (using the irredundancy of the
set $h_1,\ldots,h_d$) that there are no other automorphisms. (For details,
cf. [Ba 74c] or [Lo 79, Ch. 12. Problem 6].) □

    One can easily guess the idea behind the above construction: the
colors of the Cayley diagram are in some way encoded between and on

the two floors of the graph. (in most cases, ω can actually be omitted.
We only need it when the induced subgraphs on both floors have same
valence.) This approach naturally leads to O(nd) edges but it is not
clear at all whether this much is indeed necessary. O(n) may well be
enough, although this seems very difficult to prove. At any rate, I
find this may be the *central problem* in the topic of this section.
EDGE-MINIMUM PROBLEM 5.20. *Does* e(G) = O(n) *hold?*

In other words, does there exist a constant c such that every
finite group G can be represented by a graph with no more than cn edges
(n = |G|)?

A positive answer would follow if one could construct graphs
representing G with a bounded number of *edge-orbits*. My guess here is,
however, that such graphs don't exist. (The fact that even if they
exist, they cannot have semiregular automorphism groups, follows
immediately from the Contraction Lemma 3.7.)

The next problem seems interesting for its likely outcome will be
a *lower bound* rather than a construction.
EDGE-ORBIT CONJECTURE 5.21. *For any constant C, the following is false:*
Every finite group G can be represented by a graph with at most C edge-
orbits. (i.e., the action of G partitions the edge set into no more
than C orbits.)

The infinite version of 5.21 seems interesting as well.
EDGE-ORBIT CONJECTURE 5.22 (infinite case). *For any cardinal κ the*
following is false: Every group can be represented by a graph with at
most κ edge-orbits.

5.21 and 5.22 are closely related to vertex-orbit problems for
digraphs with bounded out-degrees.

Let X be a digraph representing G, with out-degrees not exceeding
k. Let orb(G,$\vec{k}$) denote the minimum number of orbits of Aut X for such
digraphs X. 5.21 is equivalent to this:
CONJECTURE 5.23. *For every fixed k, the numbers* orb(G,$\vec{k}$) *are unbounded*
(G ranges through all finite groups).

Similarly, the infinite version is equivalent to 5.22.

The edge-orbit conjecture is equivalent to its edge-coloured,
directed graph version. It is also equivalent to its analogue for
hypergraphs with bounded rank. The *rank* of a hypergraph is the maximum
size of its hyperedges. As soon as the rank is allowed to grow as O(d),
a bounded number of edge orbits will suffice: 7 orbits are enough if
the rank bound is 3d. This follows easily from the proof of 5.19, using
that every vertex except ω has valence <3d in the graph constructed
there.

The following is a particular case of an instability result from
the author's dissertation.

THEOREM 5.24. *Given any pair of groups* G_1, G_2 *there exists a graph* X *and an edge* e *of* X *such that* Aut $X \cong G_1$ *and* Aut$(X\backslash e) \cong G_2$.

(A more general form of this result is stated and applied in [Ba 78d].) A proof of 5.24 can be found in [Lo 79, Ch. 12, Probl. 11]. The graph constructed has $|V(X)| = O(n_1 n_2)$ vertices ($n_i = |G_i|$). It would be desirable to have a $O(n_1 n_2)$ lower bound under fairly general conditions. The problem boils down to the following question on permutation groups, which we state in a more general form, for the case of several groups.

Let Ω be a set. By a coloring of Ω we mean a function $f:\Omega \to K$ where K is a set of colors. For G a group acting on K, the subgroup Aut$_G(f)$ is defined to consist of those $g \in G$ satisfying $f(gx) = f(x)$ for all $x \in \Omega$. We call a family of colorings $f_1, \ldots, f_k$ *nearly equal*, if there is an $\omega \in \Omega$ such that $f_1(x) = \ldots = f_k(x)$ for all $x \in \Omega \backslash \{\omega\}$.

PROBLEM 5.24. *Let* Ω *be a set and* $f_1, \ldots, f_k$ *a family of nearly equal colorings of* Ω. *Let* G *be a group acting on* Ω *and suppose that the groups* G_i = Aut$_G(f_i)$ *have pairwise coprime orders. Does it follow that*

$$|\Omega| \geq |G_1| \cdot \ldots \cdot |G_k| ?$$

We remark that this inequality, if true, is best possible. Given $G_1, \ldots, G_k$ one can choose G to act on the set $\Omega = G_1 \times \ldots \times G_k$ as the wreath product of the regular representations of the groups $G_1, \ldots, G_k$ (in any order), and the colorings f_i can be appropriately chosen. Observe the non-uniqueness of the suspected extreme configurations. (There are k! ways of ordering $G_1, \ldots, G_k$ and the resulting groups G as well as the corresponding families of colorings will be quite different.)

Whenever a class c of graphs (lattices, etc.) is found to be f-universal, the question of minimizing the order of those members of c representing a given group arises. We conclude this section with some interesting open problems of this type.

As mentioned in Section 2, Steiner triple systems and strongly regular graphs are f-universal [Me 78a,b]. In order to prove this, E. Mendelsohn's idea was to first represent a given group G by permutations of a basis of a projective space V over GF(2) regarded as a Steiner triple system; and subsequently modify some of the small-dimensional subspaces of V such that the only automorphisms of the modified system will be those representing G. This approach yields a triple system with 2^{cn} vertices. The resulting strongly regular graph (cf. 2.7) will also have exponential size.

PROBLEM 5.25. *Does there exist a constant* c *such that every group* G *of order* n *is represented by* (i) *a Steiner triple system of order* $\leq n^c$; (ii) *a strongly regular graph of order* $\leq n^c$?

In these cases, even the "subgroup problem" is open.

PROBLEM 5.26. *Given a group G of order n, find a Steiner triple system X of order $\leq n^c$ on which G acts faithfully* (i.e. G is iso-morphic to some subgroup of Aut X).

Finally, let me mention some similar problems on lattices.

PROBLEM 5.27. *Does there exist a constant c such that every group of order n is represented by a lattice of order not exceeding cn?*

This would be an immediate consequence of a positive answer to the Edge-minimum Problem 5.20. As a matter of fact, *given a graph X = (V,E) there is a lattice L of order $|V| + |E| + 2$ such that Aut X $\tilde{=}$ Aut L* [Fr 50]. The proof is very simple: let L = {0,1}∪V∪E (disjoint union) and set x < y whenever x∈V, Y∈E and x is incident with y. In addition, set 0 < x < 1 for X∈V∪E. This partial order defines a lattice, satisfying the conditions.

COROLLARY 5.28 [Fr 50]. *Every finite group is represented by a lattice of order* $\leq O(nd)$.

In fact, from 5.19 we obtain the bound $|L| \leq 2n(d+2)+3$.

The above argument shows that for a lattice L representing G, the number of orbits of Aut L may be required to be as small as O(d). We do not believe that substantial further improvement is possible here.

CONJECTURE 5.29. *For every number c there exists a finite group G such that if L is a lattice representing G then Aut L has more than c orbits.*

Of course, if this is true it implies the Edge-orbit Conjecture 5.21. One can also ask the infinite version of 5.29 (c an infinite cardinal, G an infinite group).

For distributive lattices, the situation is essentially clear [Ba 80b] (the ideas go back to [Bi 45]). Every finite group G can be represented by a distributive lattice of order less than 2^{3n}, and there is an exponential lower bound $|L| > 2^p$ on the order of those distributive lattices L admitting an automorphism of prime order p. Still, the problem of representing groups without large prime divisors remains open.

PROBLEM 5.30. *For every c, find a 2-group G that cannot be represented by distributive lattices of order* $\leq n^c$.

For modular lattices, all we have is the above exponential upper bound (since every distributive lattice is modular) and no lower bound at all. Any solution to the following problem is likely to require some worthwhile contribution to the combinatorial theory of modular lattices.

PROBLEM 5.31. *For every c, find a finite group G that cannot be represented by modular lattices of order* $\leq n^c$.

6. VERTEX TRANSITIVE REPRESENTATION

A graph X is vertex-transitive if Aut X acts transitively on
V(X). After our efforts in Section 5 to minimize the number of orbits
of Aut X, it is quite natural to ask:

QUESTION 6.1. *Which groups can be represented by vertex-transitive
graphs* (digraphs, tournaments, Steiner triple systems, etc.)?

This question does not belong to the abstract representation
problem treated so far. It is a problem of *concrete representation;
which permutation groups G on a set V coincide with* Aut X *for some
graph* X = (V,E)?

In contrast to the abstract representation problem, the concrete
representation problem is intractable in this generality. The obvious
necessary condition that G be 2-closed is far from sufficient.

Regular permutation groups are an important class of 2-closed
transitive permutation groups. They are not too interesting for a
group theorist: every group G is isomorphic to (up to equivalence)
exactly one regular permutation group, the (left) regular representation
of G (the group of left translations, cf. Section 1). Even for this
class of permutation groups, the concrete representation problem turned
out to be quite difficult.

A *graphical* (digraphical) *regular representation* (GRR, resp. DRR)
of a group G is a graph (digraph) X such that Aut X is a regular
permutation group isomorphic to G.

The existence problem of GRR's and DRR's has been extensively
studied. Let us start with DRR's. Here, the result is complete:

DRR EXISTENCE THEOREM 6.2. *With five exceptions, every group (both
finite and infinite) admits a DRR.* The exceptions are the elementary
abelian groups of orders 4, 8, 9, 16 and the quaternion group of order
8.

For the finite case, the proof [Ba 80b] is a complicated version
of the proof of 5.19. It operates on an irredundant set of generators.
Such sets do not exist in general for infinite groups, a fact that
makes the proof for the infinite case substantially more difficult
[Ba 78c]. As a substitute for an irredundant generating set, one needs
a large subset of a group which is free of the relation $x^{-1}y = y^{-1}z$.
Let us call a subset H of a group G *good*, if for any x,y,z∈H, the
relation $x^{-1}y = y^{-1}z$ implies x=z.

PROBLEM 6.3. *Given a cardinal* κ, *does there exist a group of power* κ
without good subsets of the same cardinality?

Let us call such groups *strange*.

There are no countable strange groups [Ba 78c, HN 75, Su 75], and
more generally, no strange groups of weakly compact cardinality
[Ba 78c, Lemma 3.2]. (A cardinal κ is *weakly compact* if the following

Ramsey-type theorem holds: Every graph of cardinality κ contains
either a complete or an empty induced subgraph of the same cardinality.)
My guess is that strange groups exist for all successor cardinals, in
particular for $\kappa = \omega_1$. Constructing such a group may turn out to be
very difficult.

Abelian groups are not strange [HN 75, Gu 75]; in fact, they have
a good generating set [BI 79].

In our search for large good subsets of a group, *Ramsey-type*
theorems (Erdös-Rado partition calculus [EHMR]) are helpful. (For
more references, cf. [Ba 78c].) The key to the application of these
tools is provided by the following simple lemma [Ba 78c, Sublemma 3.3].
For G a group, let B(G) denote the set of *bad triples*:
$B(G) = \{\{x,y,z\} : x,y,z \in G, x^{-1}y = y^{-1}z, x \neq z\}$. For $g \in G$, let
$X_g = (G\backslash\{x\}, B_g)$ be the graph defined by $B_g = \{[x,y] : \{x,y,g\} \in B(G)\}$.
LEMMA 6.4. *The graph* X_g *contains no* K_7 (complete subgraph on 7
vertices).

It follows that the 3-uniform hypergraph (G,B(G)) contains no
complete 8-tuple, i.e. there is no 8-subset in G all triples of which
were bad.

By Ramsey-type arguments of Erdös and Rado it follows that any
"large" subset K of G contains a "large enough" good subset L, since
"good" means precisely that L contains no triple from B(G).

These observations do not yield an answer to Problem 6.3, but
combined with arguments of more technical nature they suffice for
the needs of a proof of Theorem 6.2. There will be another application of Lemma 6.4 to subgraphs of Cayley graphs in Section 8. There
we give more details of what "large enough" means for finite groups
in the above argument.

An *oriented* graph is a digraph (V,E) where $E \cap E^{-1} = \emptyset$.
PROBLEM 6.5. *Which groups admit a regular representation by an*
oriented graph?

Let us call a group G a *generalized dihedral* group if G contains
a subgroup A of index 2 and an element $g \in G\backslash A$ such that $g^2 = 1$ and
$g^{-1}ag = a^{-1}$ for all $a \in A$.

Clearly, A is an abelian normal subgroup of G and any element
$g \in G\backslash A$ satisfies the same relations. Generalized dihedral groups are
precisely those groups not generated by their elements of order greater
than 2 [Ba 78c, Prop. 5.2]. Therefore, these groups have no oriented
Cayley digraphs at all. It seems likely that almost all other groups
admit an oriented DRR.
CONJECTURE 6.6. *All groups except generalized dihedral groups and a*
finite number of other finite groups admit an oriented DRR.

This is certainly true for finite groups of odd order.

THEOREM 6.7 [BI 79]. *Every finite group of odd order has a regular representation by a tournament.* There is one exception: the elementary abelian group of order 9.

The proof employs an extension procedure relying on the Feit-Thompson theorem. It would be desirable to have a proof avoiding this reference. Otherwise there is little hope to extend the result to infinite 2'-groups (groups without elements of order 2). We remark that for infinite 2'-groups G having a *good generating set*, there is a tournament T representing G such that Aut T has *two orbits* only [BI 79]. In the general case (no good generating sets) we don't know whether a finite bound on the number of orbits can be achieved.

Other connections between DRR's and the minimization of the number of orbits for members of various universal classes are discussed in [Ba 78c] and [Ba 80b]. One of the immediate consequences of the DRR Existence Theorem 6.2 is the following:

COROLLARY 6.8. *With five possible exceptions, every group G is represented by a nilpotent semigroup S(G) such that* Aut S(G) *has only 3 orbits.*

The possible exceptions are those groups mentioned in Theorem 6.2.

□ The proof is very simple. Let X = (V,E) be a DRR of the group G. Set S = V∪{0,δ} where 0,δ are special symbols. Define xy = 0 for all x,y∈S except when x,y∈V and (x,y)∈E. For such pairs set xy = δ.

It is clear that S possesses the required properties. □ Note that Aut S is nearly a regular permutation group: the two other orbits are singletons. Also observe that S satisfies the identity xyz = 0 and products in S take only two values. As R.I. Tyshkevich commented in a talk, there is good reason to consider these the simplest kind of nontrivial semigroups.

In contrast with 6.8, one can prove that groups usually don't admit a transitive representation by a semigroup:

THEOREM 6.9 [BP 78]. *If a semigroup S admits a transitive torsion group action, then S is a rectangular band* (direct product of a left-zero and of a right-zero semigroups).

There are partial results on the two-orbit case as well. Those semigroups S which admit a torsion group G ≤ Aut S acting transitively on S\{s$_o$} for some s$_o$∈S have been classified [BP 78]

In contrast to semigroups, *every infinite group has a regular representation by a quasigroup* [BF 78]. This again follows from the DRR Existence Theorem. The methods of [BF 78] fail for the finite case.

PROBLEM 6.10 *Which finite groups admit a transitive representation by a quasigroup?*

The group of order 2 does not. Probably there is an infinity of

exceptions. The combinatorialist's interest in this problem derives from the fact that Latin squares are precisely the multiplication tables of quasigroups hence we are talking about a kind of nice symmetrical Latin squares. - Even the subgroup problem is open here:
PROBLEM 6.11. *Which transitive (regular) permutation groups are subgroups of* Aut Q *for some quasigroup* Q?

The *graphical regular representation* problem has a fairly long history. Already Frucht noticed that, *if* H *is a subset of the group* G *invariant under a group automorphism* α∈Aut G *then* α *is an automorphism of the Cayley graph* X = Γ(G,H) [Fr 52, Thm. 2.4]. In particular X is not a GRR unless α is the identity (since α certainly fixes the identity of G). We arrive at a necessary condition for the existence of GRR's.
GRR obstacle 6.12. *For all symmetric subsets* H^{-1} = H⊆G *there exists a nonidentity automorphism of the group* G *leaving* H *invariant.*

Clearly, if G satisfies 6.12, it has no GRR [No 68, Wa 71]. Watkins proposed the conjecture that the converse also holds: Every group has a GRR except for those prevented from having one by the obstacle 6.12 [Wa 71].

As a result of the combined effort of several researchers, this conjecture has been proved for finite groups.

Abelian groups of exponent ≥3 fall under the obstacle [Sa 64, Ch. 64]. The elementary abelian 2-groups of orders different from 4, 8, 16 have GRR's [Im 69]. Extension methods introduced by Nowitz and Watkins [NW 72] and further developed by Imrich [Im 76], [IW 74] culminated in the work of Hetzel [He 76] determining all *solvable* groups without a GRR. Finally G.D. Godsil proved with much ingenuity, that every non-solvable finite group admits a GRR [Go 80, Go 81].

In order to gain a better understanding of the obstacle, let us interchange the order of the two quantifiers in 6.12. We obtain a smaller class C of groups: G *belongs to* C *iff* G *has a non-identity automorphism* α *such that* αg∈{g,g^{-1}} *for all* g∈G.

Since all groups in C satisfy the condition 6.12, none of them has a GRR. The content of the Hetzel-Godsil theorem is that there are only 13 further finite groups, each of order ≤32, without a GRR.

The groups in class C can be characterized. The group G is called a *generalized dicyclic* group, if G has a subgroup A of index 2 and an element g∈G\A of order 4 such that $g^{-1}ag = a^{-1}$ for all a∈G. (Clearly, A is an abelian normal subgroup and every element of G\A has order 4 and satisfies the same relation.)
THEOREM 6.13 [Wa 71]. *For a (finite or infinite) group* G, *the following are equivalent:*

(i) G *belongs to the class* C;

(ii) G *is either a generalized dicyclic group or an abelian group of exponent* ≥3.

The main result follows:
The Hetzel-Godsil Theorem 6.14 ([He 76] + [Go 81]). *For a finite*

group G, the following are equivalent:

(i) G has no GRR;

(ii) G is either a generalized dicyclic group or an abelian group of exponent ≥3 or one of 13 exceptional groups.

The exceptions are listed in [Go 81]. Unfortunately, Hetzel's work has not been published.

It is somewhat surprising how difficult 6.14 is. Given a group G not listed under (ii) one has to find a symmetrical subset $H^{-1} = H \subseteq G$ such that the Cayley graph $\Gamma(G,H)$ has no non-identity automorphisms fixing $1 \in G$. Is it difficult to find such a set because there aren't many of them? This is hard to believe. Godsil, Imrich, Lovasz and the author have made the following conjecture.

CONJECTURE 6.15. *Let G be a finite group which is neither abelian nor generalized dicyclic. Select a symmetric subset $H^{-1} = H \subseteq G$ at random. The probability that the obtained random Cayley graph $\Gamma(G,H)$ is a GRR for G tends to 1 while n = |G| approaches infinity.*

Of course, H should not be invariant under any non-identity automorphism of G so we had better start with proving that this is indeed almost always the case. The following handy lemma is a stronger version of 6.13.

LEMMA 6.16 [Ba 78e]. *Let G be a group of order n and $D \subseteq G$ a subset of size |D| < n/8. Assume that there exists a non-identity automorphism α of G such that $\alpha g \in \{g, g^{-1}\}$ for each $g \in G \backslash D$. Then G is either a generalized dicyclic group or an abelian group of exponent ≥3.*

From this, we derive the following

COROLLARY 6.17. *Let G be a group of order n. Assume G is neither abelian nor generalized dicyclic. The probability that a random symmetric subset $H^{-1} = H \subseteq G$ is invariant under some non-identity automorphism of G is less than*

$$\frac{|Aut \ G|}{2^{n/32}} < 2^{-n/32 + \log_2^2 n} .$$

□ As a matter of fact, let k be the number of solutions of $x^2 = 1$ in G. Then m = (n+k)/2 is the number of pairs $\{g, g^{-1}\}$, and 2^m is the number of symmetric subsets of G. For a non-identity $\alpha \in Aut \ G$, let $F(\alpha) = \{g \in G : \alpha(g) \in \{g, g^{-1}\}\}$. By 6.16, $|F(\alpha)| \leq 7n/8$ hence the number of pairs $\{g, g^{-1}\}$ in $F(\alpha)$ does not exceed $(|F(\alpha)|+k)/2 \leq m-(n/16)$. Let $f(\alpha)$ denote the number of orbits of the action of α on the set of pairs $\{g, g^{-1}\}$. We have $f(\alpha) \leq m-n/32$. On the other hand, the number of symmetric sets invariant under α is $2^{f(\alpha)}$. We conclude that the probability in question does not exceed

$$2^{-m} \sum_\alpha 2^{f(\alpha)} \leq (|Aut \ G| - 1)2^{-n/32}.$$

Finally, we note that $|Aut \ G| < n^{\log_2 n}$ since G is generated by no more than $\log_2 n$ elements. □

A similar but simpler argument proves

PROPOSITION 6.18. *Let G be any group of order* n. *The probability that a random subset* H⊆G *is invariant under some non-identity automorphism of G is less than* |Aut G|$2^{-n/4}$.

This is related to the digraph version of Conjecture 6.15: with no restrictions on G, almost all Cayley digraphs should be DRR's.

The first steps toward 6.15 have been done by Godsil [Go 81b]. In particular, he proves 6.15 for nonabelian p-groups with no homomorphism *onto* $\mathbb{Z}_p$ wr $\mathbb{Z}_p$, the Sylow p-subgroup of the symmetric group of degree p^2.

We note that the countable analogues of 6.18 and 6.17 are interesting problems. A random subset of a countable set is obtained by selecting each element independently with probability 1/2.

PROBLEM 6.19. *Prove or disprove: For a countably infinite group G, the probability that a random subset of* G *is invariant under some automorphism of* G *is equal to zero.*

We have not discussed so far how the regular representation problems relate to Question 6.1, the problem of vertex-transitive representation.

Let us call a group G *dull* if all faithful transitive permutation representations of G are regular. It is easy to see that G is dull iff every nontrivial subgroup of G contains a non-trivial normal subgroup of G. For finite groups this is equivalent to saying that all subgroups of prime order in G are normal. I have just been told by P.P. Pálfy, that:

PROPOSITION 6.20. *Finite dull groups are solvable.*

The structure of dull groups may be quite complex. Huppert quotes unpublished notes of Thompson regarding dull p-groups [Hu 67, p. 342, III/12.2]. Dull groups are, perhaps, an exciting area for research.

For dull groups, 6.1 is equivalent to the corresponding regular representation problem. Clearly, abelian as well as generalized dicyclic groups are dull [No 68]. Therefore, apart from the 13 exceptional groups, the Hetzel-Godsil Theorem 6.14 gives an answer to the transitive representation problem as well. The dihedral groups D_3, D_4, D_5 have to be removed from the list of exceptions since they admit vertex-transitive representations. Five groups on the list are dull and therefore retain their exceptional status, and probably the remaining five groups have no transitive representation either.

As for DRR's: all the five exceptions in 6.2 are dull, so they don't, all other groups, do admit vertex-transitive representations by digraphs.

7. ENDOMORPHISMS

The abstract representation problem can be raised in the more general context of endomorphisms.

Let $X = (V,E)$ and $Y = (W,F)$ be (di)graphs. A map $f:V \to W$ is an $X \to Y$ *homomorphism* if $(f(v_1),f(v_2)) \in F$ whenever $(v_1,v_2) \in E$. For example, a graph X has a homomorphism into the complete graph K_s if and only if X is s-colorable in the sense of chromatic graph theory.

An *endomorphism* of X is an $X \to X$ homomorphism. Endomorphisms form a *monoid* (semigroup with identity), denoted by End X. It is natural to ask, which (abstract) monoids are represented as endomorphism monoids of graphs. The answer is as simple as for automorphism groups (2.1, 2.2), although the proof is more involved (particularly in the infinite case). The fundamental result is due to Pultr, Hedrlin and Vopenka [Pu 64, PH 64, VPH 65, HP 65, HP 66].

THEOREM 7.1. *Given a monoid M there exists a graph X such that* End $X \cong M$. *For finite monoids X can be chosen to be finite.*

A particularly clear exposition of the proof is given in [HL 69].

☐We shall outline the proof for the finite case and attempt to minimize the size of the resulting graph (Cor. 7.6).

The basic idea is Frucht's: first find a colored directed graph (endomorphisms preserve colors by definition) and then get rid of the colors.

The first step is achieved by the *Cayley diagram of the monoid*. Let M be a monoid. Define a colored digraph X on the vertex set M by joining x to xy by an edge of color y, for all $x \in M$, $y \in M \setminus \{1\}$ where 1 is the identity of M. Clearly, all left translations of M will preserve these colors and therefore M is a submonoid of End X. In order to see that $M \cong$ End X we prove that all endomorphisms of X are left translations. Indeed, for $\alpha \in$ End X we assert that α is the left translation by $z = \alpha(1)$. For $y \in M$ there is an edge of color y from 1 to y. This implies the presence of an edge of the same color from $z = \alpha(1)$ to $\alpha(y)$. The unique such edge leaving z ends at zy, hence $\alpha(y) = zy$ as asserted.

Next we show how to get rid of the colors in a relatively economic way, i.e. without a too drastic increase in the number of vertices. We derive the endomorphism analogue of Prop. 5.8.

First we have to note that the stipulation that the colors be disjoint, made before 5.8, is not appropriate here. Already the Cayley diagram of a monoid (see above) has pairs of vertices joined by several colors. This unfortunate circumstance is due to the absence of the cancellation law: $xy_1 = xy_2$ does not imply $y_1 = y_2$. This is why we can't get an upper bound better than $O(n^{3/2})$ for the

number of vertices of a graph representing a given monoid of order
n (Cor. 7.6).

In the first step, we replace our edge-colored digraph by a
vertex-colored graph where each vertex has exactly one color.

PROPOSITION 7.2. *Given an edge-colored digraph* X *with* n *vertices
and* s *(not necessarily disjoint) edge-colors, there exists a
vertex-colored graph* Y *such that*

(i)*the vertices of* Y *are partitioned into at most* $1+2\lceil\sqrt{s}\rceil$
color-classes of n *vertices each;*

(ii)*End* Y $\cong$ *End* X.

$\square$Let m = $\lceil\sqrt{s}\rceil$. We prove 7.2 by constructing a graph Y on the
vertex set $V\times\{1,\ldots,2m+1\}$ where $V = V(X)$. The vertices on the "i^{th}
floor" $V\times\{i\}$ will have color i. The vertices of the $(2m+1)^{st}$ floor
serve for "vertical identification": there will be an edge joining
(x,i) to $(x,2m+1)$ for all $x\epsilon V$, $i=1,\ldots,2m$. Let f be an injective
function

$$f : \{1,\ldots,s\} \rightarrow \{1,\ldots,m\}\times\{1,\ldots,m\}.$$

The edge-color i will be encoded as follows: for (x,y) an edge
of color i in X, we connect (x,j_1) to $(y,m+j2)$ by an edge, where
$(j_1,j_2) = f(i)$.

It should be clear now, that the endomorphisms of Y are precisely
the mappings of the form $(x,i) \rightarrow (\alpha x,i)((x,i)\epsilon V(Y)))$ where $\alpha\epsilon$End X.$\square$

REMARK 7.3. For disjoint colors, a much better estimate can be
obtained. Let $X = (V;E_1,\ldots,E_k)$ be a k-colored digraph such that
the colors $E_i\subseteq V\times V$ are pairwise disjoint. Let s denote the smallest
integer such that $k \le \binom{s}{\lceil s/2\rceil}$. Let h be an injective map from

$\{1,\ldots,k\}$ into the power set of $\{1,\ldots,s\}$ such that the sets
$h(1),\ldots,h(k)$ form a clutter, i.e. $h(i)\subseteq h(j)$ implies i=j. Let X'
denote the colored digraph $X' = (V;R_1,\ldots,R_s)$ where $R_j = \cup\{E_i : j\epsilon h(i)\}$

It is easy to verify that End X' = End X. Now, an application
of 7.2 gives the following result.

PROPOSITION 7.4. *Given an edge-colored digraph* X *with* k *disjoint
edge-colors, there exists a vertex-colored graph* Y *such that*

(i) *the vertices of* Y *are partitioned into at most*
$2+2\lceil\sqrt{\log_2 k}\rceil$ *color-classes of size* $|V(X)|$ *each;*

(ii) *End* Y $\cong$ *End* X.

In the case of automorphisms, the range of h does not need to be
a clutter. Therefore by the slightly weaker inequality $k \le 2^s-1$ we
obtain $1+2\lceil\sqrt{\log_2(k+1)}\rceil$ vertex colors. From this, Prop. 5.8 is
immediate. Vertex colors can be replaced by adjacencies to vertices
of a suitable asymmetric (automorphism free) auxiliary graph.

Endomorphisms can be dealt with in a similar fashion although there

are some nontrivial technical details to be considered. Just to
mention one: in order to avoid endomorphisms which would map the
auxiliary graph A into the original graph Y, one has to choose the
chromatic number of A to exceed that of Y. (This is the main
difficulty in proving the result that every monoid is represented
by graphs with a given subgraph [HCKN 71], [BN 78].) Observe that
the (vertex-colored) graph Y constructed in the proof of 7.2 is 3-
chromatic. (It becomes bipartite on removing the $(2m+1)^{st}$ floor.)
This makes it easy to derive the following bound.

PROPOSITION 7.5. *Given an edge-colored digraph with* n *vertices and*
s *(not necessarily disjoint) edge colors, there exists a graph* Y *such*
that End X $\tilde{=}$ End Y *and* $|V(Y)| \leq (n+1)(1+2^{\lceil \sqrt{s} \rceil})+C$. C *is an absolute*
constant, $C \leq 20$.) □

COROLLARY 7.6. *Every monoid of order* n *is isomorphic to* End X *for*
some graph X *having at most* $(2+o(1)n^{3/2}$ *vertices.*

In fact our estimate is $O(n\sqrt{d})$ where d is the minimum number of
generators of the monoid. Unfortunately, however, d may be as large
as n-1. Nevertheless, the $O(n^{3/2})$ estimate seems very poor.
J. Nesetril and the author have raised the following problem.

PROBLEM 7.7. *Does there exist a constant* C *such that every monoid of*
order n *is isomorphic to* End X *for some graph* X *having at most* Cn
vertices?

It would be particularly interesting to find a non-linear lower
bound for an infinite class of monoids.

Leaving the subject of minimization, we note that the remarks at
the end of Section 2 apply to endomorphisms as well as to automorphisms:
a very large variety of classes of combinatorial, algebraic and topo-
logical structures has been found to be universal with respect to
endomorphism monoids. The reader is referred to the nice monograph by
Pultr and Trnkova [PT 80] for more details, references and further
generalizations. Here I would only like to mention some examples.
Any monoid is isomorphic to the endomorphism monoid of an integral
domain [FS 77] and in fact of a *unique factorization domain,* as shown
very elegantly by J. Kollár [Ko 78]; of a *2-unary algebra* [HP 66];
of a *lattice* with 0 and 1 (endomorphisms preserve 0 and 1 by definition)
[GS 70]; of a *graph containing an arbitrary prescribed subgraph*
[HCKN 71], [BN 78]. (Remarkably, the proof of the last result for
infinite subgraphs is based on a theorem of Erdös and Hajnal [EH 66]
saying that there exist graphs of arbitrarily large (infinite)
chromatic number without short odd cycles.)

There are many more similar universality type results and few
non-trivial results on non-universality. No non-trivial group is the
(0-1 preserving) endomorphism monoid of a distributive lattice with 0

and 1, since every distributive lattice has (several) homomorphisms onto {0,1}. Adams and Sichler observed that the nilpotent cyclic monoid $S = \{1,a,a^2,\ldots,a^n = 0\}$ of order n+1 is no submonoid of the endomorphism monoid of any lattice of height n [AS 77].

Unary algebras are edge-colored graphs and therefore deserve the graph theorist's attention. A k-unary algebra is a k-edge-colored graph with each vertex having outdegree one in each color. An idempotent is a one-element subalgebra, i.e. a vertex carrying a loop in each color. I believed that 2-unary algebras containing a given subalgebra A without idempotents represented any monoid. I asked J. Kollár, then a first year undergraduate, to prove this. He soon came up with a disproof and gave a full characterization of the counterexamples.

THEOREM 7.8 [Ko 79, 80]. *There is a 2-unary algebra B of cardinality $2^{\aleph_0}$ without idempotents such that for any 2-unary algebra A the following are equivalent:*

    (i) *Every monoid is isomorphic to* End X *for some 2-unary algebra X containing A.*

    (ii) *A has neither loops, nor B as a subalgebra.*

We state the crucial lemma.

LEMMA 7.9 [Ko 79]. *Every 2-unary algebra without idempotents has a factor algebra of cardinality no more than $2^{\aleph_0}$ without idempotents.*

The proof of this lemma is pure combinatorics. It relies on a theorem of Erdös and Rado on delta-systems of infinite sets [ER 60].

The results generalize to k-unary algebras for any k ≥ 2. For infinite k, one has to replace $2^{\aleph_0}$ by 2^k.

Kollár has obtained a similar characterization for all varieties of 2-unary algebras with idempotent operations [Ko 81].

Note that distributive lattices, lattices of height 3 and 2-unary algebras with a given subalgebra are universal with respect to automorphism groups. For finite graphs, a similar contrast is provided by graphs of bounded valence. Not every finite monoid is represented by graphs with bounded valences [BP 80]. In fact, much more is true.

THEOREM 7.10 [BP 80]. *Given a finite graph Y there exists a finite monoid M such that any graph X satisfying* End X $\cong$ M *contains a subdivision of Y.*

This result provokes questions on what classes of monoids M force the graphs X to contain a subdivision of any given graph.

PROBLEM 7.11. *Does 7.10 remain valid if M is restricted to be an idempotent monoid? Or even a commutative idempotent monoid (semilattice)?*

Actually, we don't even know whether these classes of monoids can be represented by graphs of *bounded valence*.

Closing Remark 7.12. Let me finally emphasize again that non-trivial
new non-universality type results would be most interesting for they
would provide the first steps towards a theory establishing connections
rather than independence of the structure of combinatorial and
algebraic objects and of their abstract endomorphism monoids. Lower
bounds on the size of such objects representing a given monoid serve
similar purposes.

REFERENCES

[AS 77] M.E. Adams and J. Sichler, Bounded endomorphisms of lattices
    of finite height, Canad. J. Math. 29 (1977) 1254-1263.

[ABS 80] M.E. Adams, L. Babai and J. Sichler, Automorphism groups of
    finite distributive lattices with a given sublattice of fixed
    points, Monatsh. Math. 90 (1980) 259-266.

[Ba 72] L. Babai, Automorphism groups of planar graphs I, Discrete
    Math. 2 (1972), 285-307.

[Ba 73]   -   , Groups of graphs on given surfaces, Acta Math.
    Acad. Sci. Hung. (1973), 215-221.

[Ba 74a]   -   , Automorphism groups of graphs and edge-contraction,
    Discrete Math. 8 (1974), 13-20.

[Ba 74b]   -   , A remark on contraction of graphs with given group,
    Acta Math. Acad. Sci. Hung. 25 (1974), 89-91.

[Ba 74c]   -   , On the minimum order of graphs with given group,
    Canad. Math. Bull. 17 (1974), 467-470.

[Ba 75]   -   , Automorphism groups of planar graphs II, in:
    Infinite and finite sets (Proc. Conf. Keszthely, Hungary, 1973,
    A. Hajnal et al. eds.) Bolyai - North-Holland (1975), 29-84.

[Ba 77]   -   , Some applications of graph contractions, J. of
    Graph Theory 1 (1977), 125-130.

[Ba 78a]   -   , Chromatic number and subgraphs of Cayley graphs,
    in: Theory and Appl. of Graphs (Y. Alavi and D.R. Lick, eds.)
    Springer Lecture Notes in Math. vol. 642 (1978), 10-22.

[Ba 78b]   -   , Embedding graphs in Cayley graphs, in: Probl.
    Combinatoires et Theorie des Graphes (Proc. Conf. Paris-Orsay
    1976, J-C. Bermond et al., eds.) Centre National de Rech. Sci.,
    Paris (1978), 13-15.

[Ba 78c]   -   , Infinite digraphs with given regular automorphism
    groups, J. Combinatorial Theory - B 25 (1978), 26-46.

[Ba 78d]   -   , Vector representable matroids of given rank with
    given automorphism group, Discrete Math. 24 (1978), 119-125.

[Ba 78e]   -   , On a conjecture of M.E. Watkins on graphical
    regular representations of finite groups, Compositio Math. 37
    (1978), 291-296.

[Ba 80a]   -   , On the complexity of canonical labelling of
    strongly regular graphs, SIAM J. on Computing 9 (1980), 212-216.

[Ba 80b]   -   , Finite digraphs with given regular automorphism
    groups, Periodica Math. Hung. 11 (1980), 257-270.

[Ba 80c]   -   , Isomorphism testing and symmetry of graphs, in:
    Combinatorics 79 (M. Deza and I.G. Rosenberg, eds.), Annals of
    Discrete Math. 8 (1980), 101-109.

36

[Ba 81] L. Babai, On the order of uniprimitive permutation groups, Annals of Mathematics 113 (1981).

[Ba y]   -   , BIBD's with given automorphism groups, in preparation.

[Ba z]   -   , Recent progress in isomorphism testing, in preparation.

[BD 81] L. Babai and D. Duffus, Dimension and automorphism groups of lattices, Alg. Universalis 11 (1981).

[BF 78] L. Babai and P. Frankl, Infinite quasigroups with given regular automorphism groups, Alg. Universalis 8 (1978), 310-319.

[BG] L. Babai and C.D. Godsil, Automorphism groups of random Cayley graphs, in preparation.

[BI 79] L. Babai and W. Imrich, Tournaments with given regular group, Aequationes Math. 19 (1979), 232-244.

[BL 73] L. Babai and L. Lovász, Permutation groups and almost regular graphs, Studia Sci. Math. Hung. 8 (1973), 141-150.

[BP 78] L. Babai and F. Pastijn, On semigroups with high symmetry, Simon Stevin 52 (1978), 73-84.

[BP 80] L. Babai and A. Pultr, Endomorphism monoids and topological subgraphs of graphs, J. Comb. Theory - B 28 (1980) 278-283.

[Bi 74] N.L. Biggs, Algebraic Graph Theory, Cambridge Univ. Press, Cambridge 1974.

[BW 79] N.L. Biggs and A.T.White, Permutation groups and Combinatorial Structures, London Math. Soc. Lecture Note 33, Cambridge Univ. Press 1979.

[Bi 45] G. Birkhoff, Sobre los grupos de automorfismos, Revista Union Math. Argentina 11 (1945), 155-157.

[Bo 79] B. Bollobás, Graph Theory, Springer, N.Y. 1979.

[BS 52] R.C. Bose and T. Shimamoto, Classification and analysis of partially balanced incomplete block designs with two associate classes, J. Amer. Statist. Assoc. 47 (1952), 151-184.

[Bu] J.M.J. Buczak, quoted in [Ca 81].

[Bu 11] W. Burnside, The theory of groups of finite order (2nd ed.) 1911.

[Ca 80] P.J. Cameron, 6-transitive graphs, J. Comb. Th.-B 28 (1980), 168-179.

[Ca 81]   -   , Permutation groups and finite simple groups, Bull. London Math. Soc. to appear.

[Ca x]   -   , Automorphism groups of graphs, in: Selected Topics in Graph Theory II. (L.W. Beineke and R.J. Wilson, eds.) to appear.

[CGS 78] P.J. Cameron, J-M. Goethals and J.J. Seidel, Strongly regular graphs having strongly regular subconstituents, J. Algebra 55 (1978), 257-280.

[Ca 1878a] A. Cayley, On the theory of groups, Proc. London Math. Soc. 9 (1878), 126-133.

[Ca 1878b]   -   , The theory of groups: graphical representations, Amer. J. Math. 1 (1878), 174-176.

[Ch 64] C-Y. Chao, On a theorem of Sabidussi, Proc. A.M.S. 15 (1964), 291-294.

[Co 50] H.S.M. Coxeter, Self-dual configurations and regular maps, Bull. Amer. Math. Soc. 56 (1950), 413-455.

[CM 57] H.S.M. Coxeter and W.O.J. Moser, Generators and relations for discrete groups, Springer, Berlin 1957; 3rd ed. 1972.

[CKS 76] C.W. Curtis, W. Kantor and G. Seitz, The 2-transitive permutation representations of the finite Chevalley groups, Trans. A.M.S. 218 (1976), 1-57.

[De 73] P. Delsarte, An algebraic approach to association schemes of coding theory, Philips Res. Repts. Suppl. 10 (1973).

[De 68] P. Dembowski, Finite Geometry, Springer 1968.

[De 73] M. Deza, Une propriété extrémale des plans projectifs finis dans une classe de codes equidistants, Discrete Math. 6 (1973), 343-352.

[EH 66] P. Erdös and A. Hajnal, On chromatic number of graphs and set-systems, Acta Math. Acad. Sci. Hung. 17 (1966), 61-99.

[EH 72] - , On Ramsey-like theorems, problems and results, in: Combinatorics (Proc. Conf. Comb. Math., Oxford 1972), Inst. Math. Appl., Southend-on-Sea 1972, pp. 123-140.

[DHMR] P. Erdös, A. Hajnal, A.,Máté and R. Rado, Combinatorial Set Theory: The Ordinary Partition Relation, Akadémiai Kiadó, Budapest, to appear.

[ER 60] P. Erdös and R. Rado, Intersection theorems for systems of sets I, J. London Math. Soc. 35 (1960), 85-90.

[FH 64] W. Feit and G. Higman, The non-existence of certain generalized polygons, J. Algebra 1 (1964), 114-138.

[FK 79] E. Fried and J. Kollár, Automorphism groups of algebraic number fields, Math. Zeitschr. 163 (1978), 121-123.

[FK 81] - , Automorphism groups of fields, in: Universal Algebra (Proc. Conf., Esztergom 1977, E.T. Schmidt et al. eds.), Coll. Math. Soc. J. Bolyai 24 (1981), to appear.

[FS 77] E. Fried and J. Sichler, Homomorphisms of integral domains of characteristic zero, Trans. A.M.S. 225 (1977), 163-182.

[Fr 38] R. Frucht, Herstellung von Graphen mit vorgegebener abstrakter Gruppe, Compositio Math. 6 (1938), 239-250.

[Fr 49] - , Graphs of degree 3 with a given abstract group, Canad. J. Math. 1 (1949), 365-378.

[Fr 50] - , Lattices with a given abstract group of automorphisms, Canad. J. Math. 2 (1950), 417-419.

[Fr 52] - , A one-regular graph of degree three, Canad. J. Math. 4 (1952), 240-247.

[Go 80] C.D. Godsil, Neighbourhoods of transitive graphs and GRR's. J. Comb. Th.-B 29 (1980), 116-140.

[Go 81a] - , GRR's for non-solvable groups, in: Algebraic Methods in Combinatorics (Proc. Conf. Szeged, L. Lovasz et al. eds.), Coll. Math. Soc. J. Bolyai 25 (1981), North-Holland, to appear.

[Go 81b] - , On the full automorphism group of a graph, Combinatorica 1 (1981), to appear.

[Gnd 78] Ya. Gol'fand, k-regular graphs, private communication.

[Gr 68] G. Grätzer, Universal Algebra, Van Nostrand, Princeton 1968.

[Gr 78] - , General Lattice Theory, Birkhäuser, Basel 1978.

[GS 70] G. Grätzer and J. Sichler, On the endomorphism semigroup (and category) of bounded lattices, Pacific J. Math. 35 (1970), 639-647.

[Gr 58] J. de Groot, Automorphism groups of rings (Abstract),
Internat. Congr. Math. Edinburgh (1958), p. 18.

[Gr 59] - , Groups represented by homeomorphism groups I,
Math. Annalen 138 (1959), 80-102.

[Gu 75] R.G. Gurevich (Leningrad), private communication.

[Ha 43] M. Hall, Projective planes, Trans. A.M.S. 54 (1943), 229-277.

[Ha 69] F. Harary, Graph Theory, Addison-Wesley, Reading, Mass. 1969.

[HL 69] Z. Hedrlin and J. Lambek, How comprehensive is the category
of semigroups? J. Algebra 11 (1969), 195-212.

[HP 65] Z. Hedrlin and A. Pultr, Symmetric relations (undirected
graphs) with given semigroups, Monatsh. Math. 69 (1965), 318-322.

[HP 66] - , On full embeddings of categories
of algebras, Illinois J. Math. 10 (1966), 392-406.

[HL 74] H. Heineken and H. Liebeck, The occurrence of finite groups
in the automorphism group of nilpotent groups of class 2, Arch.
Math. 25 (1974), 8-16.

[HCKN 71] P. Hell, V. Chvátal, L. Kucera and J. Nesetril, Every
finite graph is a full subgraph of a rigid graph, J. Comb. Th.
11 (1971), 284-286.

[He 67] C. Hering, Eine Bemerkung über Automorphismengruppen von
endlichen projektiven Ebenen und Möbiusebenen, Arch. Math. 18
(1967), 107-110.

[He 72] - , On 2-groups operating on projective planes,
Illinois J. Math. 16 (1972), 581-595.

[He 76] D. Hetzel, Über reguläre graphische Darstellungen von
auflösbaren Gruppen, Diplomarbeit, Technische Universität Berlin,
1976.

[HN 75] J. Hickman and B.H. Beumann, A question of Babai on groups,
Bull. Austral. Math. Soc. 13 (1975), 355-368.

[Hi 75a] D.G. Higman, Invariant relations, coherent configurations
and generalized polygons, in: Combinatorics (M. Hall and J.H. van
Lint, eds.), Math. Centre Amsterdam 1975, 347-363.

[Hi 75b] - , Coherent configurations I, II, Geometriae Ded. 4
(1975), 1-32. and 5 (1976), 413-424.

[HS 60] A.J. Hoffman and R.R. Singleton, On Moore graphs of diameters
2 and 3, IBM J. Res. Dev. 4 (1960), 497-504.

[Hu 67] B. Huppert, Endliche Gruppen I, Springer, Berlin 1967.

[Im 69] W. Imrich, Graphs with transitive abelian automorphism
group, in: Combinatorial Th. and Appl. (P. Erdös et al. eds.),
Coll. Soc. J. Bolyai 4, North-Holland 1969, 651-656.

[Im 76] - , Graphical regular representations of groups of
odd order, in: Combinatorics (A. Hajnal and V.T. Sós, eds.), Coll.
Math. Soc. J. Bolyai 18, North-Holland (1976), 611-621.

[IW 74] W. Imrich and M.E. Watkins, On graphical regular representation
of cyclic extensions of groups, Pac. J. Math. 55 (1974), 461-477.

[Iz 57] H. Izbicki, Reguläre Graphen 3., 4. und 5. Grades mit
vorgegebenen abstrakten Automorphismengruppen, Farbenzahl und
Zusammenhängen, Monatsh. Math. 61 (1957), 42-50.

[Iz 60] - , Reguläre Graphen beliebigen Grades mit
vorgegebenen Eigenschaften, Monatsh. Math. 64 (1960), 15-21.

[Jó 72] B. Jónsson, Topics in Universal Algebra, Springer Lecture
Notes 250, Springer 1972.

[J 1869] C. Jordan, Sur les assemblages de lignes, J. Reine Angew. Math. 70 (1869), 185-190.

[Ka 74] W.M. Kantor, 2-transitive designs, in: Combinatorics (M. Hall and J.H. van Lint, eds.), Math. Centrum, Amsterdam 1974, 44-97.

[Ko 78] J. Kollár, Some subcategories of integral domains, J. Algebra 54 (1978), 329-331.

[Ko 79,80] - , The category of unary algebras containing a given subalgebra, Acta Math. Acad. Sci. Hung. 33 (1979), 407-417 and 35 (1980), 53-57.

[Ko 81] - , The category of idempotent 2-unary algebras containing a given subalgebra, in: Universal Algebra (Proc. Conf., Esztergom 1977, E.T. Schmidt et al., eds.), Coll. Math. Soc. J. Bolyai 24, North-Holland (1981).

[Kö 36] D. König, Theorie der endlichen und unendlichen Graphen, Akad. Verlag, Leipzig 1936.

[Lo 79] L. Lovász, Combinatorial Problems and Exercises, Akadémiai Kiadó, Budapest 1979.

[Lu 80] E.M. Luks, Isomorphism of bounded valence can be tested in polynomial time, in: Proc. 21st IEEE FOCS Symposium, IEEE, Long Beach, Calif. 1980, 42-49.

[Ma 68a] W. Mader, Homomorphiesätze für Graphen, Math. Ann. 175 (1968), 154-168.

[Ma 68b] - , Über trennende Eckenmengen in homomorphiekritischen Graphen, Math. Ann. 175 (1968), 243-252.

[Ma 72] - , Wohlquasigeordnete Klassen endlicher Graphen, J. Comb. Th. 12 (1972), 105-122.

[Ma 1896] H. Maschke, The representation of finite groups, especially of the rotation groups of the regular bodies in three- and four-dimensional space, by Cayley's Color Diagrams, Amer. J. Math. 18 (1896), 156-194.

[Me 72] E. Mendelsohn, Every group is the collineation group of some projective plane, J. Geometry 2 (1972), 1-9.

[Me 78a] - , On the groups of automorphisms of Steiner triple and quadruple systems, J. Comb. Th.-A 25 (1978), 97-104.

[Me 78b] - , Every (finite) group is the group of automorphisms of a (finite) strongly regular graph, Ars Combinatoria 6 (1978), 75-86.

[MQ 79] D.J. McCarthy and L.V. Quintas, The construction of minimal-line graphs with given automorphism group, in: Topics in Graph Theory (F. Harary, ed.) Ann. N.Y. Acad. Sci. 328 (1979), 144-156.

[Mo 64] J.W. Moon, Tournaments with a given automorphism group, Canad. J. Math. 16 (1964), 485-489.

[Ne 77] P.M. Neumann, Finite permutation groups, edge-coloured graphs and matrices, in: Topics in Group Theory and Computation (M.P.J. Curran, ed.) Acad. Press, London 1977, 82-118.

[No 68] L.A. Nowitz, On the non-existence of graphs with transitive generalized dicyclic groups, J. Comb. Th. 4 (1968), 49-51.

[NW 72] L.A. Nowitz and M.E. Watkins, Graphical regular representations of non-abelian groups I-II, Canad. J. Math. 24 (1972), 993-1008 and 1009-1018.

[O 62] O. Ore, Theory of graphs, A.M.S. Colloq. Publ. 38, Providence R.I. 1962.

[Po 37] G. Pólya, Kombinatorische Anzahlbestimmungen für Gruppen, Graphen und chemische Verbindungen, Acta Math. 68 (1937), 145-254.

[Pu 64] A. Pultr, Concerning universal categories, Comment, Math. Univ. Carolinae 5 (1964), 227-239.

[PH 64] A. Pultr and Z. Hedrlin, Relations (graphs) with given infinite semigroups, Monatsh. Math. (1964), 425-445.

[PT 80] A. Pultr and Vera Trnková, Combinatorial, Algebraic and Topological Representations of Groups, Semigroups and Categories, Academia Praha, Prague 1980.

[Sa 57] G. Sabidussi, Graphs with given automorphism group and given graph theoretical properties, Canad. J. Math. 9 (1957), 515-525.

[Sa 59]   -   , On the minimum order of graphs with given auto-morphism group, Monatsh. Math. 63 (1959), 124-127.

[Sa 60]   -   , Graphs with given infinite group, Monatsh. Math. 64 (1960), 446-457.

[Sa 64]   -   , Vertex-transitive graphs, Monatsh. Math. 68 (1964), 426-438.

[ST 81] J.J. Seidel and D.E. Taylor, Two-graphs, a second survey, in: Algebraic Methods in Graph Theory (L. Lovász and V.T. Sós, eds.), Coll. Math. Soc. J. Bolyai 25 (1981), to appear.

[Sh 54] I.R. Shafarevich, Construction of fields of algebraic numbers with given solvable Galois group, Izvestiya Akad. Nauk SSSR Ser. Math. 18 (1954), 525-578.

[Su 75] A. Souslin (Leningrad), private communication.

[Ti 74] J. Tits, Buildings of spherical type, Springer Lecture Notes 1974.

[VPH 65] P. Vopenka, A. Pultr and Z. Hedrlin, A rigid relation exists on any set, Comment. Math. Univ. Carolinae 6 (1965), 149-155.

[Wa 71] M.E. Watkins, On the action of non-abelian groups on graphs, J. Comb. Th.-B 1 (1971), 95-104.

[Wa 74]   -   , Graphical regular representations of alternating, symmetric and miscellaneous small groups, Aequat. Math. 11 (1974), 40-50.

[We 81] R. Weiss, A note on s-transitive graphs, Combinatorica 1 (1981), to appear.

[Wi 64] H. Wielandt, Finite permutation groups, Acad. Press, N.Y. 1964.

[Wi 69]   -   , Permutation groups through invariant relations, Lecture notes, Ohio State University 1969.

[Wil] R.M. Wilson, private communication (1980).

[Wh 73] A.T. White, Graphs, Groups and Surfaces, North-Holland, Amsterdam 1973.

A TOUR THROUGH TOURNAMENTS

OR

BIPARTITE AND ORDINARY TOURNAMENTS: A COMPARATIVE SURVEY

LOWELL W. BEINEKE
PURDUE UNIVERSITY AT FORT WAYNE AND
THE POLYTECHNIC OF NORTH LONDON

Some of the richest theory in the study of directed graphs is
found in the area of tournaments, and it is interesting to see whether
a portion of that theory can be applied to other areas. In this work,
we will examine analogues of results on 'ordinary' tournaments in
another family, the 'bipartite' tournaments.

We begin by defining these two families. An *ordinary tournament*
is a directed graph with the property that every pair of vertices are
joined by exactly one arc; it is thus the result of orienting the
edges of a complete graph. Analogously, we call the result of
orienting the edges of a complete bipartite graph a *bipartite tournament*; formally, it is a directed graph having its vertices partitioned
into two sets with no vertices in the same set joined by an arc, and
every pair of vertices in different sets joined by exactly one arc.
In Figure 1, we show the four ordinary tournaments of order 4, and in
Figure 2, the four bipartite tournaments having two vertices in each
partite set.

Figure 1

(a)   (b)   (c)   (d)

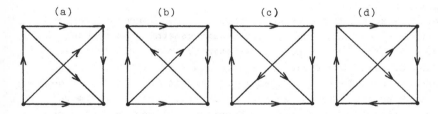

Figure 2

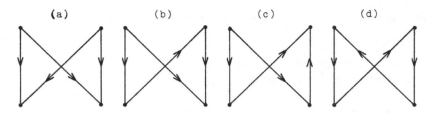

(a)   (b)   (c)   (d)

Ordinary tournaments were first studied in the context of
comparisons: for each pair of objects in a group, one is rated
superior to the other. This may be, for example, the result of
competition between teams, of pecking between chickens, or of
preference between items. There is a similar interpretation for bipartite tournaments; they can be considered as the result of competition between the individuals on two teams or comparisons between the
items in two sets.

In this survey, we will concentrate on two areas common to the
two types of tournament, cycles and scores. Our investigations into
bipartite tournaments are still in the early stages, and we hope that
this work will stimulate further research on the subject. Our format
will be to discuss a result on ordinary tournaments and then follow
this with a similar or contrasting result on bipartite tournaments.
For convenience in noting results of each type, we shall use single
vertical lines alongside theorems on ordinary tournaments and double
lines by bipartite results.

There are some references which should be mentioned before we
proceed further, and here the two areas have something in common: in
each case the primary reference was written by Moon. His doctoral
thesis [12] contains the first systematic study of bipartite tournaments, and his book Topics on Tournaments [14] is the basic work on
ordinary tournaments. Two other surveys in this area are the more
elementary article [8] by Harary and Moser and the more recent work
[16] by Reid and Beineke.

1: DEFINITIONS AND TERMINOLOGY
In the theory of tournaments there are certain concepts which are
not in general use in the rest of graph theory, and in this section we
present some of the notation and terminology which we shall be using.

The vertex-set of a tournament of either type will be denoted by
V (we shall not have occasion to use notation for the arc-set). For

bipartite tournaments, we shall generally assume that the bipartition
of V is ordered, (X,Y). An ordinary tournament with n vertices will
frequently be called an n-tournament, while a bipartite tournament
with $|X|$ = m and $|Y|$ = n will be called an m x n bipartite tournament.

In a tournament, vertex v is said to *dominate* vertex w if there
is an arc directed from v to w. The number of vertices dominated by
a given vertex v is called the *score* of v, denoted s(v), and the number
dominating v the *co-score,* s'(v). The *score list* of an ordinary tournament is the collection of the scores of its vertices, a bipartite tournament has a *pair of score lists,* with the scores of the vertices in X and
Y constituting separate lists.

The *converse* T' of a tournament T is the result of reversing each
arc of T. Clearly, the co-scores of T are the scores of T', and hence
we will call the lists of scores and co-scores *dual* lists.

We next define some special types of tournaments:
(a) Reducible and irreducible tournaments. A tournament (of either
type) is called *reducible* if some non-empty proper subset of its vertex-
set has no in-coming arcs, and *irreducible* otherwise. In each of
Figures 1 and 2, only the last tournament is irreducible. It is not
difficult to show that a tournament is irreducible if and only if every
vertex can be reached from all others along directed paths. (Irreducibility is thus sometimes called strong connectedness.) An (*irreducible*)
component of a tournament is a maximal irreducible subtournament, and
every tournament has a unique decomposition of its vertex-set into
components. A component with just one vertex is called *trivial*. In
an ordinary tournament, a non-trivial component must have at least
three vertices, while in a bipartite tournament, it must have at least
two vertices from each partite set.
(b) Consistent tournaments. A tournament which has no directed cycles
is called *consistent*. For ordinary tournaments, there is only one of a
given order, and it is usually known as the *transitive* n-tournament.
For example, the first tournament in Figure 1 is the transitive 4-
tournament. In contrast, the first three 2 x 2 bipartite tournaments
in Figure 2 are all consistent, so that in general there are many
consistent m x n bipartite tournaments. Two special types are of some
interest: (i) A *unanimous* bipartite tournament is one in which all of
the vertices in one set dominate all those in the other set.
(ii) A *linear* bipartite tournament is one in which the vertices in the
two sets can be labelled v_1, v_2, ...,v_n and w_1, w_2,..., w_n in such a
way that v_i dominates w_j if and only if $j \geq i$. Up to isomorphism
(including interchanging the sets of vertices), there is only one n x n
bipartite tournament which is unanimous and one which is linear. In
Figure 2, the first is unanimous and the second linear.
(c) Regular tournaments. A tournament is called *regular* if all of its

vertices have the same score. Clearly, a regular ordinary tournament
must have odd order, while a regular bipartite tournament must have
its partite sets of the same even order. It is not difficult to show
that all non-trivial regular tournaments are irreducible.

(d) Quadric bipartite tournaments. We call a bipartite tournament
quadric if each partite set is split as $X = X_0 \cup X_1$ and $Y = Y_0 \cup Y_1$
so that vertex x in X_i dominates vertex y in Y_j if and only if $i = j$.
(See Figure 3.) If the subsets of X have orders m_0 and m_1 and those
of Y orders n_0 and n_1, we shall denote this tournament as $Q(m_0, m_1; n_0, n_1)$,
with the special notation $Q_{n,n}$ when all four sets have the same order.

Figure 3

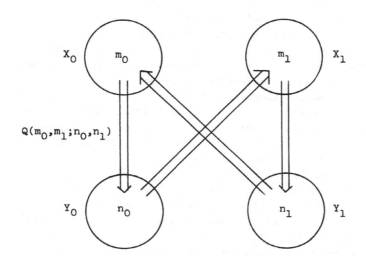

2: THE EXISTENCE OF CYCLES

We begin this section with Camion's well-known result on Hamiltonian
tournaments, which brings together the topics of cycles and irreducibility (5):

> Theorem 2.1. Every irreducible ordinary tournament has
> a Hamiltonian cycle.

In contrast to this result, the 3 x 3 example in Figure 4 shows
that not all irreducible n x n bipartite tournaments are Hamiltonian.
Of course, irreducibility implies the existence of some cycles, so it
is natural to ask for lower bounds on the longest of these. In
pursuing this question, Jackson [9] established the following result:

Theorem 2.2. Let T be an irreducible bipartite tournament
with the property that for all vertices v and w, either v
dominates w or their scores satisfy $s(v) + s'(w) \geq r$. Then
T has a cycle of length at least 2r.

Figure 4

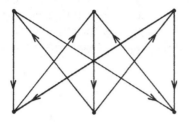

This result has two interesting corollaries as special cases:

Corollary 2.2A. In any bipartite tournament with minimum score s
and minimum co-score t, there is a cycle of length at least
$2(s+t)$.

Corollary 2.2B. Every regular bipartite tournament is Hamiltonian.

There have been several generalisations of Camion's theorem, one
of which was found by Harary and Moser [8] and is equivalent to the
following statement:

Theorem 2.3. If an ordinary tournament has an r-cycle, then
it has cycles of length 3, 4, ..., r.

Corollary 2.3A. An irreducible ordinary n-tournament has
cycles of lengths 3, 4, ..., n.

Although Camion's original result does not have a direct analogue
in bipartite tournaments, Theorem 2.3 has. Of course, no cycle in a
bipartite tournament can have odd length, and furthermore, in quadric
tournaments all cycle lengths are multiples of 4. However, as our
next result (due to Beineke and Little [3]) states, these are the
only exceptions to a pancyclic property.

Theorem 2.4. Let T be a bipartite tournament with a cycle
of length 2r. Then
(a) if r is odd, T has cycles of lengths 4, 6, ..., 2r;
(b) if r is even, T has cycles of lengths 4, 8, ..., 2r,
and unless the 2r-cycle spans $Q_{r,r}$, T has cycles of lengths
4, 6, ..., 2r.

Corollary 2.4A. Every Hamiltonian n x n bipartite tournament
except for the quadric tournament $Q_{n,n}$ (n even) has cycles of
lengths 4, 6, ..., 2n, and $Q_{n,n}$ has cycles of lengths 4, 8, ...,
2n.

3: NUMBERS OF CYCLES

In an ordinary tournament, the scores s_1, s_2, ..., s_n determine the number of 3-cycles: $\binom{n}{3} - \sum_{i=1}^{n}\binom{s_i}{2}$. They do not, however, determine the number of r-cycles for $r > 3$. For example, the two 5-tournaments in Figure 5 have the same scores, but the first has three 4-cycles while the second has four. Similarly, the two 3 x 3 bipartite tournaments in Figure 6 have the same score lists yet different numbers of 4-cycles. Therefore, we turn to bounds on the number of 4-cycles which can occur, first considering the maximum number in ordinary tournaments.

Figure 5

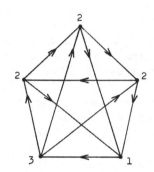 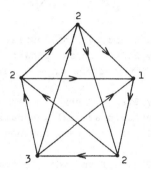

Figure 6

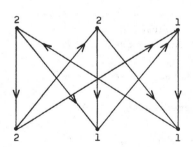 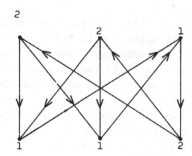

Theorem 3.1. Let $C_r(n)$ denote the maximum number of r-cycles in any ordinary n-tournament. Then

$$C_3(n) = \begin{cases} \frac{1}{24}n(n^2 - 1) & \text{for n odd,} \\ \frac{1}{24}n(n^2 - 4) & \text{for n even,} \end{cases}$$

and

$$C_4(n) = \frac{1}{2}(n-3)C_3(n) \qquad \text{for all n.}$$

The 3-cycle result follows from the fact that the expression for the number of 3-cycles given above is maximised when the scores are as nearly equal as possible. The 4-cycle result follows from a theorem of Beineke and Harary [2] giving the maximum number of irreducible subtournaments in an n-tournament. For odd values of n, the maximum occurs in the 'rotational' tournament R_{2r+1} with vertices $0, 1, \ldots, 2r$ in which i dominates $i+1, i+2, \ldots, i+r$ (mod n); for $n = 2r$, a maximising tournament is found by deleting any vertex from R_{2r+1}.

The result for bipartite tournaments corresponding to Theorem 3.1 was found by Moon and Moser [15]. In this case the maximum is realised by those quadric tournaments in which the partite sets X and Y are split as nearly equally as possible.

Theorem 3.2. The maximum number of 4-cycles in an m x n bipartite tournament is

$$C_4(m,n) = \left\lceil \frac{m}{2} \right\rceil \left\lceil \frac{m+1}{2} \right\rceil \left\lceil \frac{n}{2} \right\rceil \left\lceil \frac{n+1}{2} \right\rceil.$$

For cycles of length greater than 4, exact results remain elusive; see Moon [14] for a discussion of bounds for ordinary tournaments. The situation for Hamiltonian cycles is especially interesting. Moser (see [14]) showed that the expected number of Hamiltonian cycles in a random n-tournament is approximately $\frac{\pi}{e} (\frac{n}{2e})^{n-1}$, but no one seems to have been able to find n-tournaments with as many Hamiltonian cycles as this. We fare somewhat better with bipartite tournaments. Moon [12] showed that the expected number of Hamiltonian cycles in an n x n bipartite tournament is $\frac{(n!)^2}{n2^{2n}}$. On the other hand, for n even, the quadric tournament $Q_{n,n}$ has $\frac{2}{n}(\frac{n}{2}!)^4$ Hamiltonian cycles, and this is about πn times the expected number. He conjectures that this is in fact the maximum.

A variation of the 4-cycle problem has been shown to have a rather unexpected application. An n x n matrix H whose entries are +1 or -1 is called a *Hadamard* matrix if its rows are mutually orthogonal, that is, if $HH^T = nI$. The Hadamard Conjecture states that if n is a multiple of 4, then there exists a Hadamard matrix of order n. Little and Thuente[11] discovered a connection between this conjecture and 'odd orientations of 4-circuits':

Theorem 3.3. For n a multiple of 4, there exists a
Hadamard matrix of order n if and only if the maximum
number of linear 2 x 2 subtournaments (Figure 2(b)) in
an n x n bipartite tournament is $n^2(n-1)(n-2)/8$.

We now turn to the opposite extreme, minimising the number of
r-cycles, where we restrict our attention to irreducible tournaments
since some tournaments have no cycles. In the ordinary case, Moon
[13] found the minimum for all r; the extremal tournament is obtained
from the transitive tournament by reversing the arcs of the spanning
path.

Theorem 3.4. The minimum number of r-cycles in an
irreducible ordinary n-tournament is

$c_r(n) = n - r + 1$.

In the bipartite case, the situation is again somewhat different.
For, the quadric tournament Q(1,m-1;1,n-1) is irreducible yet has
only 4-cycles (they number (m-1)(n-1)).

Theorem 3.5. The minimum number of 4-cycles in an irreducible
m x n bipartite tournament is

$c_4(m,n) = m + n - 3$ $(m,n \geq 2)$.

4: REALISATIONS OF SCORE LISTS

The main topic in this section is the problem of determining
what collections of integers constitute the scores of tournaments.
For ordinary tournaments, an existence solution to this problem was
first found by Landau [10]; included in his proof is a reduction step
which also provides a constructive solution. We present this in a
general form first:

Theorem 4.1. Let L be a list of n non-negative integers not
exceeding n-1, and let L' be obtained from L by deleting
one entry s_k and reducing $n-1-s_k$ largest entries by 1.
Then L is the score list of some ordinary tournament if
and only if L' is.

If we begin with our list in non-decreasing order, and then
delete the final entry and maintain the non-decreasing property while
reducing entries, repeating this procedure results in a canonical
tournament with the given list as scores (if it is realisable). We
illustrate this with a simple case in Figure 7.

Figure 7.

$L_1 = [1,1,3,3,3,4]$ $\qquad$ $v_6 \rightarrow v_1, v_2, v_4, v_5$

$L_2 = [1,1,2,3,3,]$ $\qquad$ $v_5 \rightarrow v_1, v_2, v_3$

$L_3 = [1,1,2,2]$ $\qquad$ $v_4 \rightarrow v_1, v_2$ (and v_5)

$L_4 = [1,1,1]$ $\qquad$ $v_3 \rightarrow v_2$ (and v_4, v_6)

$L_5 = [0,1]$ $\qquad$ $v_2 \rightarrow v_1$, ; $v_1 \rightarrow v_3$

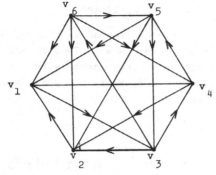

We next state Landau's existence criterion; it includes a determination of which lists belong to irreducible tournaments.

Theorem 4.2. A non-decreasing list $[s_1, s_2, \ldots, s_n]$ of non-negative integers is the score list of some ordinary tournament if and only if, for $k = 1, 2, \ldots, n$,

$$\sum_{i=1}^{k} s_i \geq \binom{k}{2}$$

with equality when $k = n$.

Furthermore, the tournament is reducible if and only if equality holds for some $k < n$.

It has been observed (Eggleton and Beineke independently, unpublished) that one need only verify the inequality for those values of k for which $s_k < s_{k+1}$ (and of course the final inequality). Thus, it follows that the list in Figure 7 is known to belong to some tournament since $1 + 1 > \binom{2}{2}$, $1 + 1 + 3 + 3 + 3 > \binom{5}{2}$ and

$$1 + 1 + 3 + 3 + 3 + 4 = \binom{6}{2},$$

and any realising tournament must be irreducible.

A bipartite tournament with partite sets X and Y gives rise in a natural way to two bipartite graphs with the same bipartition, one having edges corresponding to the arcs directed from X to Y, the other those from Y to X. It follows that results on valency lists of bipartite graphs can be applied to bipartite tournaments, and vice versa. Thus, Gale's reduction criterion [7] for valencies in bipartite graphs provides an analogue to Theorem 4.1. In what follows, A and B will be lists of non-negative integers, $A = [a_1, a_2, \ldots, a_m]$ and $B = [b_1, b_2, \ldots, b_n]$ such that each $a_i \leq n$ and each $b_j \leq m$.

Theorem 4.3. Two lists A and B in non-decreasing order are the score lists of some bipartite tournament if and only if the lists $A' = [a_1, a_2, \ldots, a_{m-1}]$ and $B' = [b_1, \ldots, b_{a_m}, b_{a_m+1}-1, \ldots, b_n-1]$ are.

If, in this theorem, the list B' is formed so that the non-decreasing property is preserved, then, as with Theorem 4.1, we have a construction for canonical bipartite tournaments with given score lists. This is illustrated in Figure 8, where the graph shows only the arcs of the tournament from X to Y.

Figure 8

$A_1 = [1,1,3,4,4]$	$B_1 = [2,2,2,3,3]$	$x_5 \rightarrow y_1,y_2,y_3,y_5$
$A_2 = [1,1,3,4]$	$B_2 = [2,2,2,2,3]$	$x_4 \rightarrow y_1,y_2,y_3,y_4$
$A_3 = [1,1,3]$	$B_3 = [2,2,2,2,2]$	$x_3 \rightarrow y_3,y_4,y_5$
$A_4 = [1,1]$	$B_4 = [1,1,2,2,2]$	$x_2 \rightarrow y_2$
$A_5 = [1]$	$B_5 = [0,1,1,1,1]$	$x_1 \rightarrow y_1$

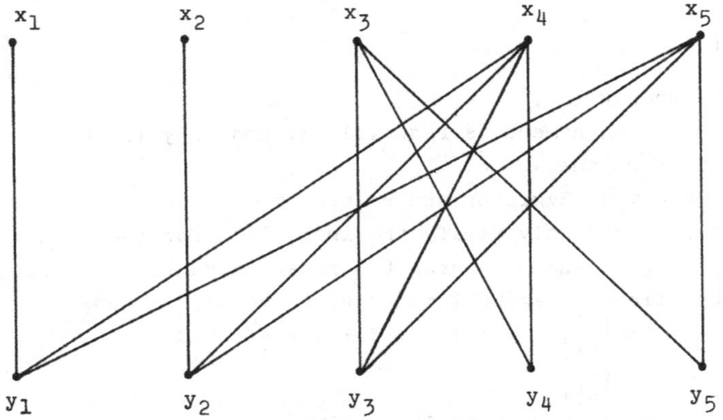

We have two bipartite analogues to Landau's existence theorem; the first due to Moon [12] and the second to Ryser [18] (and originally stated in terms of row-sums in (0,1)-matrices). (Note that the order of the entries in the lists in the two theorems are reversed for ease of stating.)

> Theorem 4.4. Two lists A and B in non-decreasing order
> are the score lists of some bipartite tournament if and
> only if, for k = 1, 2, ..., m and ℓ = 1, 2, ..., n,
>
> $$\sum_{i=1}^{k} a_i + \sum_{j=1}^{\ell} b_j \geq k\ell$$
>
> with equality when k = m and ℓ = n.
> Furthermore, the bipartite tournament is reducible if and
> only if equality holds for some lesser combination of k
> and ℓ.

> Theorem 4.5. Two lists A and B in non-increasing order
> are the score lists of some bipartite tournament if and
> only if, for k = 1, 2, ..., m,
>
> $$\sum_{i=1}^{k} a_i \leq \sum_{j=1}^{n} \min (k, m-b_j)$$
>
> with equality when k = m.
> Furthermore, the bipartite tournament is reducible if and
> only if $b_1 = m$, $b_n = 0$, or equality holds for some k < m.

As with Landau's result, only those inequalities for which there
are jumps in the lists need to be checked. We illustrate this for
the lists in Figure 8. For Moon's criterion, we form the partial
sums $A_2 = 1 + 1$, $A_3 = A_2 + 3$, $A_5 = A_3 + 4 + 4$; $B_3 = 2 + 2 + 2$,
$B_5 = B_3 + 3 + 3$, and then check that

$A_2 + B_3 = 8 > 2.3$, $A_3 + B_3 = 11 > 3.3$, $A_5 + B_3 = 19 > 5.3$

$A_2 + B_5 = 14 > 2.5$, $A_3 + B_5 = 17 > 3.5$, $A_5 + B_5 = 25 = 5.5$

For Ryser's criterion, we reverse the order in the lists and check
that

$4 + 4 < 2 + 2 + 2 + 2 + 2$, $4 + 4 + 3 < 2 + 2 + 3 + 3 + 3$, and

$4 + 4 + 3 + 1 + 1 = 2 + 2 + 3 + 3 + 3$.

In both cases, we also see that any realisation of the lists must be
irreducible.

If the scores in a tournament are neither too small nor too large,
then irreducibility must follow, a fact which we make precise in our
next two results. The first is due to Moser (see [8]), the second to
Moon (see [12] or [4]); both are best possible.

> Theorem 4.6. An ordinary n-tournament in which each
> score s_i satisfies $\frac{1}{4}(n-1) \leq s_i \leq \frac{3}{4}(n-1)$ is irreducible.

> Theorem 4.7. An m x n bipartite tournament in which
> each score a_i satisfies $\frac{n}{4} < a_i < \frac{3}{4}n$ and each score b_j
> satisfies $\frac{m}{4} < b_j < \frac{3}{4}m$ is irreducible.

To conclude this section on scores of tournaments in general, we present two results on tournaments having the same scores. These results are due respectively to Ryser [17] and to Beineke and Moon [4].

> Theorem 4.8. If two ordinary tournaments have the same scores, then each can be obtained from the other by successively reversing the arcs of 3-cycles.

> Theorem 4.9. If two bipartite tournaments have the same scores, then each can be obtained from the other by successively reversing the arcs of 4-cycles.

5: SCORES AND SPECIAL PROPERTIES

In this section we consider further topics related to scores, beginning with decompositions of reducible tournaments into components. It follows from Theorems 4.8 and 4.9 (or less directly from 4.2 and 4.4) that two tournaments with the same scores can differ only within irreducible components. Thus, the scores determine

> the number of components,
> the number of vertices in each component,
> the scores of the vertices in each component, and
> the dominance between components.

In ordinary tournaments, the inter-component dominance is complete and transitive. Hence, this structure can be determined by successively finding (and then deleting) dominating components. The following result formalises this step; it is similar to a result of Avery [1].

> Theorem 5.1. Let $L = [s_1, s_2, \ldots, s_n]$ be the score list in non-decreasing order of some reducible tournament T. Then the dominating component of T consists of those vertices with scores exceeding s_k, where k is the largest index for which k < n and
> $$\sum_{i=1}^{k} s_i = \binom{k}{2}$$

For example, we can deduce that any tournament with score list [1,1,1,3,5,5,6,7,7] must have a dominating component of order 5, then a component of order 1, and then one of order 3.

Bipartite tournaments behave in a similar fashion, but trivial components complicate the situation a little. For, two trivial components in the same partite set force the inter-component structure to be less than a complete ordering. Nonetheless, we still seek dominating components (see [4]). Here again, however, trivial components complicate matters in that they need not show themselves in Ryser's inequalities.

Theorem 5.2. Let A and B be the score lists in non-increasing order of some reducible bipartite tournament. Then

(i) if $a_1 = n$ or $b_1 = m$, there is a corresponding trivial dominating component;

(ii) otherwise, if k is the minimum index for which

$$\sum_{i=1}^{k} a_i = \sum_{j=1}^{n} \min(k, m-b_j)$$

there is a non-trivial dominating component with those scores $a_i \geq a_k$ and $b_j > m-k$.

To illustrate this result, we use the lists [4,4,4,3,3,1,1] and [6,5,2,1,1]. Since $a_1 < 5$ and $b_1 < 7$, and since equality first holds for $k = 3$, we have a component with three vertices of score 4 in one set and two vertices with scores 6 and 5 in the other. After deleting these scores, we have the lists [3,3,1,1] and [2,1,1], so there are next two trivial components from X (with score 3), followed by one from Y (with score 2). Finally, there is a 4-component (all scores 1).

From this, we turn to a determination of those lists belonging to exactly one tournament; we call such lists *simple*. Because the decomposition into irreducible components is fixed from the scores, we need only consider the irreducible case. These results were first proved by Avery [1] and Beineke and Moon [4].

Theorem 5.3. The simple score lists of irreducible ordinary tournaments are [0], [1,1,1], [1,1,2,2], and [2,2,2,2,2].

Theorem 5.4. Let A and B be the score lists of an irreducible m x n bipartite tournament with $m, n \geq 3$. Then A and B are a simple pair if and only if

(i) one is [1,1,...,1] or its dual, or

(ii) one is [d,1,...,1] with $d > 1$ or its dual and the other is constant.

We observe that Theorem 5.4 effectively determines those pairs of valency lists belonging to just one bipartite graph.

The following result of Eplett [6] answers another question on tournament scores:

Theorem 5.5. A score list $L = [s_1, s_2, \ldots, s_n]$ in non-decreasing order belongs to some self-converse ordinary tournament if and only if, for $k = 1, 2, \ldots, [\frac{1}{2}n]$,

$$s_k + s_{n-k+1} = n - 1.$$

Thus, all that is required for a list to belong to a self-converse ordinary tournament is that the scores and co-scores be the same, that is, that the list be self-dual. The natural bipartite analogue of this

would seem to be that the lists A and B be self-dual, that is,
$a_i + a_{m-i+1} = n$ and $b_j + b_{n-j+1} = m$ for all i and j. These conditions
are certainly necessary, but it so happens that they are not
sufficient. For, the score lists [1,1,1,3,3,3] and [2,2,4,4] are
self-dual, but no tournament with these lists is self-converse. Thus,
we are left without an analogue to Theorem 5.5.

To balance this lack of a result, we conclude with a theorem on
bipartite tournaments for which the analogue for ordinary tournaments
is extremely simple: The only score list of a consistent ordinary
n-tournament is [0,1,...,n-1].

> Theorem 5.6. Let A and B be lists in non-decreasing
> order and with combined sum mn. These lists belong
> to some consistent bipartite tournament if and only if
> (i) $a_m = n$, and for i = 1,2, ..., m-1, there are
> $a_{i+1} - a_i$ scores b_j equal to i, or
> (ii) $b_n = m$, and for j = 1, 2, ..., n-1, there are
> $b_{j+1} - b_j$ scores a_i equal to j.

For example, it is readily verified that the lists A = [0,2,3,4]
and B = [1,1,2,3] satisfy the conditions: $a_4 = 4$, and the consecutive
differences 2, 1, 1 in A are the numbers of occurrences of 1, 2, and
3 in B. The corresponding consistent tournament is shown in Figure 9.

Figure 9

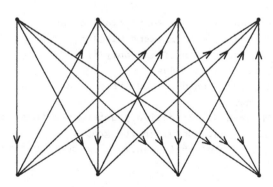

With this, we conclude our brief tour through tournaments.

REFERENCES

[1]   P. Avery, The condition for a tournament score sequence to be simple, J. Graph Theory 4(1980) 157-164.

[2]   L.W. Beineke and F. Harary, The maximum number of strongly connected subtournaments, Canad. Math. Bull. 8(1965) 491-498.

[3]   L.W. Beineke and C.H.C. Little, Cycles in complete oriented bipartite graphs, J. Combinatorial Theory (□)(to appear).

[4]   L.W. Beineke and J.W. Moon, On bipartite tournaments and scores, Proceedings of The Fourth International Conference on the Theory and Applications of Graphs, Kalamazoo, MI, May 6-9, 1980, to appear.

[5]   P. Camion, Chemins et circuits hamiltoniens des graphes complets, C.R. Acad. Sci. Paris (A) 249(1959) 2151-2152.

[6]   W.J.R. Eplett, Self-converse tournaments, Canad. Math. Bull. 22(1979) 23-27.

[7]   D. Gale, A theorem on flows in networks, Pacific J. Math. 7(1957) 1073-1082.

[8]   F. Harary and L. Moser, The theory of round robin tournaments, Amer. Math. Monthly 73(1966) 231-246.

[9]   B. Jackson, Long paths and cycles in oriented graphs, submitted.

[10] H.G. Landau, On dominance relations and the structure of animal societies: III. The condition for a score sequence, Bull. Math. Biophysics 15 (1953) 114-118.

[11] C.H.C. Little and D.J. Thuente, Hadamard matrices and bipartite graphs, J. Australian Math. Soc.(□)(to appear).

[12] J.W. Moon, On some combinatorial and probabilistic aspects of bipartite graphs, Ph.D. thesis, University of Alberta, Edmonton, 1962.

[13] J.W. Moon, On subtournaments of a tournament, Canad. Math. Bull. 9(1966) 297-301.

[14] J.W. Moon, Topics on Tournaments, Holt, Rinehart and Winston (New York), 1968.

[15] J.W. Moon and L. Moser, On the distribution of 4-cycles in random bipartite tournaments, Canad. Math. Bull. 5(1962) 5-12.

[16] K.B. Reid and L.W. Beineke, Tournaments, Chapter 7 in Selected Topics in Graph Theory (L.W. Beineke and R.J. Wilson, eds.). Academic Press (London, New York, San Francisco), 1978, pp.169-204.

[17] H.J. Ryser, Combinatorial properties of matrices of zeros and ones. Canad. J. Math. 9(1957) 371-377.

[18] H.J. Ryser, Matrices of zeros and ones in combinatorial mathematics. Recent Advances in Matrix Theory. Univ. of Wisconsin Press (Madison), 1964, pp.103-124.

Shift Register Sequences

Henry Beker and Fred Piper.

1.   Introduction

An n-stage shift register consists of n binary storage
elements $S_0, S_1, \ldots, S_{n-1}$, called <u>stages</u>, connected in series.
The contents of the stages change in time with a clock pulse
according to the following rule:- if $s_i(t)$ denotes the
content of S_i after the tth time pulse then $s_i(t+1) = s_{i+1}(t)$
for i = 0,...,n-2 and $s_{n-1}(t+1) = f(s_0(t), s_1(t), \ldots, s_{n-1}(t))$.
The function f is called the <u>feedback function</u> of the register.
If, for any t, we write $s_t = s_o(t)$ then we say that the register
generates the sequences (s_t).   Clearly $s_i = s_0(i)$ for all i
satisfying $0 \leq i \leq n-1$, and the sequence (s_t) is completely
determined by $s_0, s_1, \ldots, s_{n-1}$ and the feedback function f.

Shift registers are, for many different reasons, of
considerable interest to mathematicians.   In this paper we
will discuss one of their applications in cryptography.   We
suppose that we have some binary data which we wish to
transmit in 'disguised' form so that any interceptor will not
be able to deduce our original data.   One way to try this is
illustrated in the following diagram.

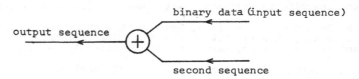

binary data (input sequence)

output sequence

second sequence

The symbol ⊕ represents a modulo 2 adder which, as the

name suggests, produces an output sequence by adding (mod 2)
the corresponding entries of the two sequences.   So, if
(a_t) is the input sequence and (b_t) is our second sequence,
the output sequence will be (a_t+b_t).   Since all addition is
modulo 2, any two of the sequences completely determine the
third.

The input sequence is determined by the data to be
transmitted.   It might, for instance, be a binary representation of a message originally written in English.   In this
case it will exhibit certain patterns which reflect the
statistics of letter frequencies etc. in the English language.
The second sequence should be chosen so that it 'hides' these
statistics as much as possible.   Consequently we require the
second sequence to be as patternless or unpredictable as
possible.   There are many statistical criteria for determining
'good' choices for this second sequence.   The final conclusion
also depends on what else you demand of your sequence.

When messages are transmitted in this way it frequently
happens that an interceptor is able to work out, (or even guess),
the input equivalent of some part of the output sequence.
(For example, it might be clear that the input was a letter which
began with the word 'Dear' and ended "Yours sincerely".)   Once
he has this knowledge the interceptor automatically knows some
part of the second sequence.   For this paper we will demand of
our second sequence that, as well as satisfying the usual
statistical requirements, a 'reasonably long' part of the
sequence does not determine the sequence completely.   (Of
course we will need to be a little more precise about the phrase

'reasonably long'.)   Given this demand the object of the
paper is to discuss ways of using shift registers as
building blocks for generating second sequences.

In section 2 we discuss the sequences generated by
shift registers with linear feedback functions. For these
the length of sequence necessary to determine the entire
sequence is unacceptably small.   In section 3 we mention,
very briefly, the use of a shift register with a non-linear
feedback function and then, in section 4, give an equally
brief discussion of one way of using more than one register.
Both these prove to be unsatisfactory and are included merely
to illustrate that, in order to satisfy our criteria, having
exceptionally long periods and/or appearing extremely complex
are not sufficient.   In section 5 we then include a rather
longer discussion of multiplexing.   This is a method of
combining two linear sequences which, so far as we can tell,
seems a reasonable building block for our second sequence.

We have not included any proofs.   The aim is merely to
show the type of results available and the kind of problems
which occur.   For details of section 2 see [1] and for those
of section 5 see [2].   For further details of J-K flip flops
see [3],[4].

2. Linear feedback shift registers

If $f(s_0(t), s_1(t), \ldots, s_{n+1}(t)) = \sum_{i=0}^{n-1} c_i s_i(t) \pmod 2$,

(with each c_i equal to 0 or 1), then the shift register is said to have <u>linear feedback</u>. The constants $c_0, c_1, \ldots, c_{n-1}$ are called the <u>feedback coefficients</u>. This can be represented by the diagram

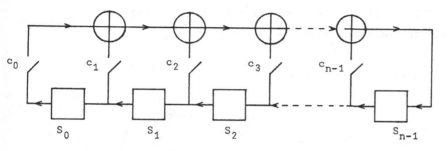

where $c_i = 1$ stands for a closed connection and $c_i = 0$ for an open one.

The content of a shift register, regarded either as a binary number or an n-bit binary vector, is called its <u>state</u>. Clearly every state has a unique successor. If all the c_i are zero then, regardless of the initial state of the shift register, after the nth clock pulse each stage S_i will contain 0 and will remain this way. Thus to keep the shift register active, at least one of the c_i must be 1. Let j be the least value of i such that $c_i = 1$, and suppose $j \geq 1$. Then the (mod 2) sum of the contents of $S_j, S_{j+1}, \ldots, S_{n-1}$ are fed back into S_{n-1}, and the contents of $S_0, S_1, \ldots, S_{j-1}$ contribute nothing to the action of the shift register. So after the jth clock pulse, the state of the shift register will be independent of $s_0(0), s_1(0), \ldots$ $s_{j-1}(0)$ and we are essentially only using n-j positions of the machine. To eliminate this possible degeneracy we will assume henceforth that j = 0, that is c_0 = 1.

If we let $\bar{s}(t)$ denote the binary vector $(s_0(t), s_1(t),$ $\ldots, s_{n-1}(t))$, then the action of the shift register may be described by the matrix equation

$$\bar{s}(t+1) = \bar{s}(t)M \quad \text{where}$$

$$M = \begin{pmatrix} 0 & 0 & 0 & \ldots & 0 & c_0 \\ 1 & 0 & 0 & \ldots & 0 & c_1 \\ 0 & 1 & 0 & \ldots & 0 & c_2 \\ \cdot & \cdot & & & \cdot & \cdot \\ \cdot & \cdot & & & \cdot & \cdot \\ 0 & 0 & & & 1 & c_{n-1} \end{pmatrix} .$$

Since we are assuming c_0 = 1, the matrix M is non-singular. This means, of course, that we can now always deduce $\bar{s}(t)$ from $\bar{s}(t+1)$ and that, apart from the initial state, each state now has a unique predecessor as well as a unique successor. Since there are only 2^n possible states, a repetition must occur among the first $2^n + 1$ states. As soon as this repetition occurs the sequence of state vectors will continue to repeat itself. Although there are 2^n possible states, the non-singularity of M guarantees that, provided the initial state is not all zeros, the all zero vector will not occur as a state. This discussion forms the basis of a proof for :

<u>Result 2.1</u> : The succession of states of an n-stage
shift register with linear feedback is periodic with period
$p \leq 2^n - 1$.

If we let $s_0, s_1, s_2, \ldots, s_{n-1}, s_n, s_{n+1}, \ldots$ denote the
successive contents of S_0, it is clear that we get an
infinite binary sequence which satisfies

$$s_{t+n} = \sum_{i=0}^{n-1} c_i s_{t+i} \quad \text{for } t = 0, 1, 2, \ldots .$$

Such an equation is called a <u>linear recurrence relation
of order n</u> and the infinite binary sequence $(s_t) =$
$s_0, s_1, s_2, \ldots$ is called a <u>linear recurring sequence</u>.

Such a sequence (s_t) may also be thought of as the
<u>output</u> of the shift register and may be said to be
<u>generated</u> by that shift register. At any given time the
state of the shift register represents a section of the
sequence. Clearly, from Result 2.1, (s_t) is periodic with
period $p \leq 2^n - 1$.

<u>Example</u> : We consider a 4-stage linear feedback shift
register with feedback coefficients $c_0 = c_3 = 1$, $c_1 = c_2 = 0$

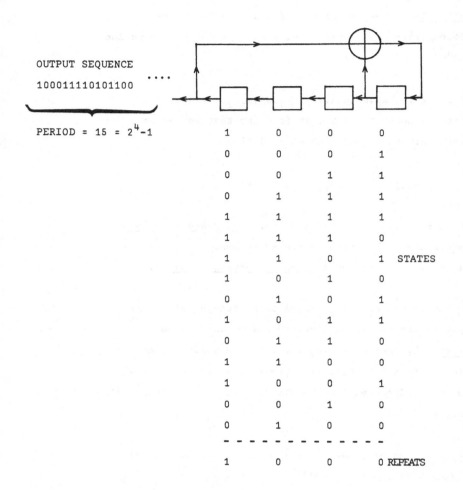

For an n-stage shift register, the 2^n different initial
states will give rise to 2^n periodic sequences.   Suppose
one such sequence has period p.   Then the first p terms

$$s_0, s_1, s_2, \ldots, s_{p-1}$$

will be called a <u>cycle of length p</u>.   The different
starting points of a cycle (p altogether) give rise to
different periodic sequences, which are all translates
of each other.   For example, the two sequences of
period 15 whose first 15 terms are given by

```
1 0 0 0 1 1 1 1 0 1 0 1 1 0 0     and
0 0 0 1 1 1 1 0 1 0 1 1 0 0 1
```

respectively, correspond to the same cycle but are
definitely different.   The second sequence is the
translate of the first by 1.

The initial state 00 .... 0 corresponds to a cycle
of length 1 and the resulting sequence is called the
<u>null sequence</u>, or <u>zero sequence</u>, denoted by (0).   The
remaining 2^n-1 sequences will be distributed among cycles
of various lengths.   If these 2^n-1 sequences all lie on
one cycle of length 2^n-1, (as is the case in our example),
then we have a <u>maximum length cycle</u> and each of these
sequences of maximum period is called a PN (pseudo-noise)
sequence.   Clearly all possible register contents except
00 .... 0 will occur once among the first 2^n-1 states of the
shift register during the generation of a PN sequence.

<u>Definition</u> :   Associated with the linear recurrence
relation of order n

$$s_{t+n} = \sum_{i=0}^{n-1} c_i s_{t+i} \qquad t = 0,1,2,\ldots$$

is the <u>characteristic polynomial f(x)</u> defined by

$$f(x) = 1 + c_1 x + c_2 x^2 + \ldots + c_{n-1} x^{n-1} + x^n$$

(remembering that $c_0 = 1$).

Consequently we may identify a shift register with the characteristic polynomial $f(x)$ and we shall refer to the sequences and cycles <u>generated</u> by $f(x)$. Of course different initial states of the shift register will give rise to different periodic sequences.

The aggregate of all infinite binary recurring sequences (s_t) generated by a given $f(x)$ is called the <u>solution space</u> of $f(x)$ and is denoted by $\Omega(f)$. Thus $\Omega(f)$ consists of 2^n sequences corresponding to all 2^n initial states. With the obvious definitions of addition and scalar multiplication, $\Omega(f)$ is an n-dimensional vector space over $GF(2)$.

There are a number of very nice results relating properties of the characteristic polynomial with the period of the sequence.

<u>Result 2.2</u> : Suppose $f(x)$ is a polynomial over $GF(2)$ with exponent e. Then the period of any sequence (s_t) in $\Omega(f)$ divides e.

<u>Result 2.3</u> : Suppose $f(x)$ is an irreducible polynomial over $GF(2)$ with exponent e. Then the period of any non-null

sequence (s_t) in $\Omega(f)$ is e.

<u>Result 2.4</u> :   Suppose $f(x)$ is a polynomial over GF(2) with
$f(0) = 1$.   Then $\Omega(f)$ contains a sequence of period 2^n-1
if, and only if, $f(x)$ is primitive.   (Note that this could
be restated as <u>every</u> non-null sequence in $\Omega(f)$ has period
2^n-1 if, and only if, $f(x)$ is primitive.)

<u>Result 2.5</u> :   Suppose $f(x)$, $g(x)$ are polynomials over GF(2)
with $f(0) = g(0) = 1$.   Then $\Omega(f) \subseteq \Omega(g)$ if, and only if,
$f(x)|g(x)$.

As a consequence of these results we know that, by
taking any primitive polynomial as our characteristic
polynomial, we will obtain a PN-sequence by choosing any
non-zero initial state.

If we have any PN-sequence with period 2^n-1 then it can
be shown that any generating cycle, i.e. any 2^n-1 consecutive
entries, contains precisely 2^n ones and 2^n-1 zeros.
Furthermore its out of phase autocorrelation is always
$\frac{-1}{2^n-1}$.   (For any binary sequence (s_t) of period p, let
A and D be, respectively, the number of argreements and
disagreements when (s_t) and (s_{t+a}) are compared in their first
p positions.   The autocorrelation function c(a) is defined
by $c(a) = \frac{A-D}{p}$.)   These facts, plus other simple statistical
tests, suggest that a PN-sequence is reasonably "pattern-free"
and suitable for use as a foundation for generating the
'second sequence' referred to in section 1.

Despite the fact that it may have period 2^n-1, any sequence generated by an n-stage shift register with linear feedback is completely determined by its characteristic polynomial f(x) and any one of its state vectors. Thus the entire sequence is known once we know the n feedback constants and any n consecutive entries of the sequence. But, as we have already seen, the feedback constants are the coefficients of a linear relation of order n which is satisfied by the sequence (s_t). This leads to :

Result 2.6 :   A PN-sequence (s_t) of period 2^n-1 is completely determined by any 2n consecutive entries.

If the known entries are $s_r, s_{r+1}, \ldots, s_{r+2n-1}$, then all that is involved in determing the feedback constants, (and hence the entire sequence), is the inversion of the non-singular matrix

$$\begin{pmatrix} s_r & \cdots & s_{r+n-1} \\ s_{r+1} & \cdots & s_{r+n} \\ & & \\ s_{r+n-1} & \cdots & s_{r+2n-1} \end{pmatrix} .$$

A routine operation unless n is very large. But there are many practical snags, (not to mention the cost!), in trying to use very large shift registers as a building block for generating the second sequence referred to in section 1. So, for our particular problem, the use of a shift register with linear feedback in not satisfactory. If we wish to use shift registers in the generation of our second sequence then we must try to remove the 'linearity' from the system.

Before we look at ways in which this has been attempted
we must make one important observation.   If (u_t) is any
binary sequence with period p then (u_t) can be generated by a
p-stage shift register with initial state $u_0, u_1, \ldots, u_{n-1}$ and
characteristic polynomial $x^p + 1$.   In other words any periodic
binary sequence is a linear shift register sequence.
However in Result 2.6 the n is the size of the shift register
and so, for any given sequence, one measure of its suitability
is the size of the smallest shift register which can generate
it with a linear feedback function.   This size is called the
linear equivalence of the sequence and what we want is to use
one (or many) "small" shift registers to obtain a sequence whose
linear equivalence is large enough so that Result 2.6 does not
bother us.

3. Non-linear feedback

Since we wish to remove the effect of a linear feedback we will now discuss the effect of using a non-linear feedback function. No matter which feedback function is used, each state will have a unique successor. Thus as soon as a state is repeated the entire sequence of states will begin to repeat itself. Since there are at most 2^n states the maximum possible period of the sequence of states is 2^n. (Note that, unlike the linear situation, the zero state need not be followed by zero.) So we certainly cannot hope to significantly increase the period of our sequence. However it is conceivable that we will considerably increase the linear equivalence. It is not difficult to show that there are 2^{2^n} possible feedback functions, for an n-stage shift register. As an illustration of one the following truth table and diagram show the effect of choosing $f(s_0,s_1,s_2) = 1+s_0+s_1+s_2+s_0s_1+s_1s_2+s_2s_0$ for a 3-stage register.

s_0	s_1	s_2	$f(s_0s_1s_2)$
0	0	0	1
0	0	1	0
0	1	0	0
0	1	1	0
1	0	0	0
1	0	1	0
1	1	0	0
1	1	1	1

Truth Table

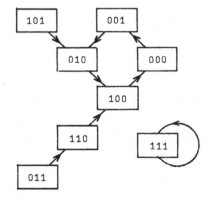

Diagram

Any state vector may be regarded as the binary representation of an integer and it is customary to order the rows of the table according to the values of these integers. All the information of the table is then in the right hand column and we will often refer to this column as the table.

There are two points to notice about this small example. Firstly it contains two cycles. Secondly there are a number of states with no predecessors, (such states are called branch points). Clearly, for any particular example, the existence of branch points reduces the size of the maximum possible period. For this reason it is usually desirable to avoid them and this is easy to do.

Result 3.1 : The state diagram of a shift register has no branch points if, and only if, the feedback function $f(s_0, s_1, \ldots, s_{n-1}) = s_0 + f'(s_1, \ldots, s_{n-1})$.

Unlike the linear situation it is not quite so easy to write down simple conditions which guarantee a sequence of maximal length. The following results give some necessary conditions.

Result 3.2 : For $n > 2$, the number of cycles in the diagram of a branchless shift register is odd if, and only if, the number of ones in the truth table is odd.

<u>Result 3.3</u> :   In order to obtain a sequence of period 2^n
from an n-stage shift register it is necessary to use all n
available tap positions in the feedback function; (i.e. each
s_i, $0 \leq i \leq n-1$, must occur in the function).

One practical consequence of Result 3.3 is that the
generation of a sequence of period 2^n requires a lot of hardware.
This becomes particularly relevant if, as is very likely, we
wish to change our feedback function.

We will not discuss non-linear feedback in as much detail
as the linear case.   The situation is summarized by saying
that non-linear feedback involves a lot more work with very
little benefit.   What is worse is that it is not always easy
to predict the outcome.   For instance the autocorrelation
function might not be good, but this is not always apparent
from looking at the function.   To partially justify these
assertions we will give an example which generates a sequence
of period 2^n and linear equivalence 2^n-1.   For the example
take $f(s_0,\ldots,s_{n-1}) = f^*(s_0,\ldots,s_{n-1}) + \bar{s}_0\bar{s}_1\ldots\bar{s}_{n-1}$ where
f^* is a linear feedback function corresponding to a primitive
polynomial and $\bar{s}_i$ is the complement of s_i in GF(2).   This
large value for the linear equivalence seems good and we have
certainly generated a sequence which would need a much larger
register with linear feedback.   Unfortunately, however, it is
still only necessary to know 2n consecutive bits to determine
the entire sequence.

4. J-K flip flops

One of the simplest ways of using more than one
shift register is to employ a J-K flip flop as a mixer.
As always we assume

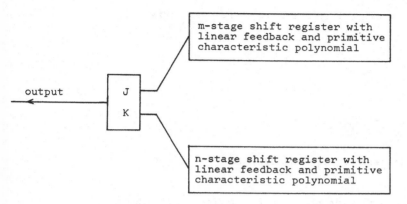

that each individual register has a linear feedback and
generates a sequence of maximal possible length. J-K flip
flops are widely used in electronic circuits and often have
more than the two inputs shown in our diagram.

The action of a J-K flip flop can probably best be
described by use of its truth table. If we let j_r, k_r and a_r
denote, respectively, the contents of J and K and the
output at time r then the truth table is

j_r	k_r	a_r
0	0	a_{r-1}
0	1	0
1	0	1
1	1	$\overline{a}_{r-1}$

This table yields the following equation :

$a_r = (j_r + k_r + 1) a_{r-1} + j_r.$ Since each output depends on the previous one we must define for the first output, i.e. for a_0 we need to know a_{-1}.   It is customary to take $a_{-1} = 0.$ If our m-stage register generates (s_t) while the n-stage register generates (u_t) then the first relevant terms of our output sequence are :

$$a_1 = s_0,$$

$$a_2 = (s_1 + u_1 + 1) s_0 + s_1,$$

$$a_3 = (s_2 + u_2 + 1) [(s_1 + u_1 + 1) s_0 + s_1] + s_2.$$

Clearly we are obtaining non-linear equations.   Furthermore the equations are becoming increasingly complex as t increases.   The sequence obtained is ultimately periodic, (i.e. periodic except for a few terms at the beginning.   More precisely, there exist integers n,p such that $a_{x+py} = a_x$ for all y and all $x \geq n$).   The period is easily calculated and, if we choose $(m,n) = 1$, can be as large as $(2^m - 1)(2^n - 1).$ However, as we shall now illustrate, knowledge of any two consecutive elements of (a_t) gives information about the individual sequences (s_t) and (u_t).   So, once again, knowledge of comparatively few consecutive elements of (a_t), (the precise number varies slightly with the circumstances), gives enough information to determine (s_t) and (u_t) and, consequently, (a_t).

Suppose that, for some particular value of r, we know

a_r and a_{r-1}. If we substitute these particular values into our truth table we obtain the following :

if $a_{r-1} = 0$ and $a_r = 0$ then $j_r = 0$

if $a_{r-1} = 0$ and $a_r = 1$ then $j_r = 1$

if $a_{r-1} = 1$ and $a_r = 0$ then $k_r = 1$

if $a_{r-1} = 1$ and $a_r = 1$ then $k_r = 0$.

But this means that a_r and a_{r-1} determine one of j_r or k_r. In other words if we know two consecutive terms of the output sequence we can deduce the value of other s_i or u_i for some i. It should now be clear why we had to be vague about the number of terms necessary to determine the entire sequence (a_t). Since knowledge of 2m consecutive entries determines (s_t) and knowledge of 2n consecutive entries determines (u_t), it is conceivable, (but highly unlikely!), that $2(m+n)+1$ consecutive entries will determine (a_t). However, in practice, the number of entries needed will not usually be much larger than this minimum value. No matter what the precise value is, it should be clear that if one knows a sufficient number of consecutive terms of (a_t) one can obtain enough information about the individual linear sequences (s_t) and (u_t) to determine them completely.

The two papers referred to discuss more sophisticated and complicated systems, based on the above ideas, for using J-K flip flops as building blocks for our second sequence.

5. <u>Multiplexing</u>

Let SR1 and SR2 be two shift registers with m
and n stages respectively, (m>1,n>1), such that each has
a linear feedback function. We denote the stages of SR1
by $A_0,\ldots,A_{m-1}$ and those of SR2 by $B_0,\ldots,B_{n-1}$.
Furthermore we let $a_i(t)$ and $b_j(t)$ denote the contents of
A_i and B_j at time t.   In order to define a multiplexed
sequence we assume that both shift registers have
primitive characteristic polynomials, i.e. that SR1 generates
a binary sequence (a_t) of period 2^m-1 and SR2 generates a
binary sequence (b_t) of period 2^n-1.   A multiplexer is a
device used to produce a sequence, which we call a
multiplexed sequence, related to the states of SR1 and SR2
in the following way. We first choose an integer k in the
range $1 \leq k \leq m$.   We can only choose k = m if $2^m-1 \leq n$ and if
k ≠ m then k must also satisfy $2^k \leq n$.   Having chosen k we
now choose k stages $a_{x_1}, a_{x_2}, \ldots, a_{x_k}$ of SR1 and, for convenience,
we assume $0 \leq x_1 < x_2 < \ldots < x_k \leq m-1$.   At any time t, the binary
k-tuple $(a_{x_1}(t), a_{x_2}(t), \ldots, a_{x_k}(t))$ is interpreted as the
binary representation of a natural number which we denote
by N_t.   Clearly, $0 \leq N_t \leq 2^k-1$ but if k = m then, since the
binary m-tuple $(0,0,\ldots,0)$ is never a state, we can improve
inequality slightly to $1 \leq N_t \leq 2^m-1$.   If k<m we choose an
injective mapping $\theta : \{0,1,\ldots,2^k-1\} \to \{0,1,\ldots,n-1\}$ while
if k = m we choose an injective mapping $\theta : \{1,2,\ldots 2^k-1\} \to
\{0,1,\ldots,n-1\}$.   (Note that the restrictions on k guarantee
the existence of such a mapping.)   With these choices of
$k,x_1,x_2,\ldots x_k$ and θ we define a new sequence (u_t), called
<u>a multiplexed sequence</u>, by $u_t = b_{\theta(N_t)}(t)$.   Basically all

that the multiplexer does is pick one of the stages of
SR2 at each time t, namely $B_{\theta(N_t)}$.   But, for any t and
any $i, b_i(t) = b_1(t+i) = b_{t+i}$ which means $u_t = b_{t+\theta(N_t)}$
for every t.

An example

The following is an illustration of a multiplexed
sequence with m = 3 and n = 4.

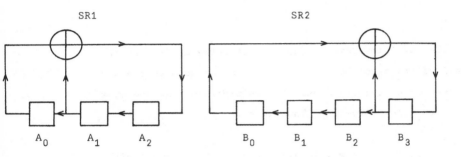

We will assume that the initial state of SR1 is 100
and that of SR2 is 1000.   For this example we take k = 2,
$x_1 = 0$, $x_2 = 1$ and let θ be the mapping $\{0,1,2,3\} \rightarrow \{0,1,2,3\}$
given by $\theta(0) = 2$ $\theta(1) = 3$ $\theta(2) = 0$ $\theta(3) = 1$.

The first seven states of each register are shown below

SR1	SR2
1 0 0	1 0 0 0
0 0 1	0 0 0 1
0 1 0	0 0 1 1
1 0 1	0 1 1 1
0 1 1	1 1 1 1
1 1 1	1 1 1 0
1 1 0	1 1 0 1

From this we see that $N_0 = 2$, $N_1 = 0$, $N_2 = 1$, $N_3 = 2$, $N_4 = 1$, $N_5 = 3$ and $N_6 = 3$. Thus $\theta(N_0) = 0$, $\theta(N_1) = 2$, $\theta(N_2) = 3$, $\theta(N_3) = 0$, $\theta(N_4) = 3$, $\theta(N_5) = 1$, $\theta(N_6) = 1$ &,finally $u_0 = b_0(0) = 1$, $u_1 = b_2(1) = 0$, $u_2 = b_3(2) = 1$, $u_3 = b_0(3) = 0$, $u_4 = b_3(4) = 1$, $u_5 = b_1(5) = 1$ and $u_6 = b_1(6) = 1$. Straightforward computation gives the first 120 terms of (u_t) as: 1 0 1 0 1 1 1 1 0 1 0 0 0 0 0 0 1 0 1 1 1 1 1 1 1 0 1

0 0 0 0 0 1 1 1 1 0 0 0 1 1 1 1 1 0 1 1 0 1 1 1 1 1 1 0 1 1

0 0 0 0 0 0 1 0 1 0 1 0 0 0 1 0 0 0 0 0 1 1 1 1 0 1 1 1 0

1 0 1 0 0 1 1 1 0 1 0 0 1 0 1 1 0 1 1 0 1 0 1 1 1 1 0 1 0 0

0 0 0.

In any multiplexed sequence each entry depends on the previous states of both registers.   So as soon as the two states repeat simultaneously the multiplexed sequence must begin to repeat itself.   Thus we have :

Result 5.1 :   The multiplexed sequence is periodic with period $p \leq (2^m-1)(2^n-1)$.

Straightforward verification shows that our example has period at least 105.   So, since $105 = (2^3-1).(2^4-1)$ its period must be 105 and so Result 5.1 is the best possible. A great deal is known about the period of multiplexed sequences and, more important, about their linear equivalences. We will restrict our attention to the situation where the multiplexed sequence is longer than that which can be obtained from either register.   For reasons which we will not even begin to explain this means we will assume $n \nmid m$.

Result 5.1 can obviously be improved to say that $p \leq$ l.c.m$(2^m-1, 2^n-1)$. The following result, whose proof is rather long, shows precisely when p is as large as possible.

<u>Result 5.2</u> : If $n \nmid m$ and $\left(2^m-1, \dfrac{2^n-1}{2^{(m,n)}-1}\right) = 1$ then the period of a multiplexed sequence is l.c.m$(2^m-1, 2^n-1)$. So, in particular, if $(m,n) = 1$ the period is $(2^m-1)(2^n-1)$.

Result 5.2 is rather powerful and a little surprising. It does not depend on either characteristic polynomial, the value of k or the choice of θ. If we write $p = 2^m-1$ and $e = 2^n-1$ then statistical analysis of a multiplexed sequence yields :

<u>Result 5.3</u> :   If $(m,n) = 1$ the mean value of the out of phase autocorrelation is $\dfrac{p-e}{e(pe-1)} \cong \dfrac{1}{e^2} - \dfrac{1}{pe}$.

Since $\dfrac{1}{e^2} - \dfrac{1}{pe}$ is small this is encouraging but, of course, the mean itself gives no information about specific values of c(a). It is desirable to have c(a) close to zero for all a in the range $1 \leq a \leq pe-1$. It is fairly straightforward to show that if n is large in comparison with m the $c(a) \cong \dfrac{-1}{2^n-1}$ for most values of a.

We are now in the situation where we can generate a sequence with a large period and with reasonable statistical properties. However we do not yet know if it is suitable as a building block for our second sequence. To decide this we need to know its linear equivalence. In general it

seems to be difficult to determine this. The best
result so far is

Result 5.4 : If $(m,n) = 1$ the linear equivalence d of a
multiplexed sequence is related to the k stages selected
from SR1 in the following way :

1. $d = n(1+m)$ if $k = 1$

2. $d = n(1+m+\binom{m}{2}))$ if $k = 2 < m$

3. $d \leq n(1+ \sum_{i=1}^{k} \binom{m}{i}))$ if $2 < k < m-1$, with equality if the

      k stages are spaced at equal intervals.

4. $d = n(2^m-1)$ if $k = m-1$ or m.

It should be noted that if the k stages are unevenly
distributed then equality need not occur in (3).   Although
Result 5.4 does not tell us d for all possible situations
it does mean that, by suitable choices of m,n and k, we
can use two linear shift registers to obtain a sequence
which appears reasonable for our purposes.

Considerable research is currently being conducted
to try to generalise and/or improve the results in this
section.   Hopefully by the time of the conference in July
there will be further progress to report.

References

[1]   E. Selmer, Linear recurrence relations over finite
            fields.   University of Bergen, Norway.

[2]   S.M. Jennings, A special class of binary sequences.
                Ph.D. Thesis, University of London, (1980).

[3]   V. Pless, Encryption schemes for computer
            confidentiality. M.I.T. Report MAC TM-63 (1975).

[4]   F. Rubin, Decrypting a stream cipher based on J-K
            flip flops.   IEEE transactions on
            computers. Vol. c-28 No. 7 (1979).

Henry Beker,   Fred Piper,
Racal Datacom Ltd.   Department of Mathematics,
Milford Industrial Estate,   Westfield College,
Tollgate Road,   (University of London)
Salisbury,   Kidderpore Avenue,
Wilts. SP1 2JG   London, NW3 7ST.

RANDOM GRAPHS
Béla Bollobás
Department of Pure Mathematics and Mathematical Statistics,
University of Cambridge.

The aim of this review is to highlight some of the fundamental
results about random graphs, mostly in areas I am particularly interested in. Though a fair number of references are given, the review is
far from complete even in the topics it covers. Furthermore, very few
of the proofs are indicated. The exception is the last section, which
concerns random regular graphs. This section contains some very recent
results and we present some proofs in a slightly simplified form.

The study of random graphs was started by Erdös [33], who applied
random graph techniques to show the existence of a graph of large
chromatic number and large girth. A little later Erdös and Rényi [38]
investigated random graphs for their own sake. They viewed a graph as
an organism that develops by acquiring more and more edges in a random
fashion. The question is at what stage of development a graph is likely
to have a given property. The main discovery of Erdös and Rényi was
that many properties appear rather suddenly. In the last twenty years
many papers have been written about random graphs; some of them, in the
vein of [33], tackle traditional graph problems by the use of random
graphs, and others, in fact the majority, study the standard invariants
of random graphs in the vein of [38]. Of course the two trends cannot
really be separated for deep applications are impossible without
detailed knowledge of random graphs. In this paper we shall be concerned mostly with the second trend but we do give occasional applications. For background information on graph theory the reader is
referred to [9]; a rich variety of random graph results can be found
in Erdös and Spencer [43], and a gentle introduction to random graphs
is in Ch.VII of [11].

Most results about random graphs concern two closely related
models. Let n be a natural number and set $N = \binom{n}{2}$. Given an integer
$M = M(n)$, $0 \leq M \leq N$, consider the set $G(n,M)$ of all $\binom{N}{M}$ graphs with a
fixed set V of n labelled vertices and with M edges. We turn $G(n,M)$
into a probability space by giving all graphs the same probability.
The elements of $G(n,M)$ are said to be *random graphs of order* n *and size*
M. We shall denote by $G_{n,M}$ an element of $G(n,M)$. In particular, given
a property Q of graphs, the probability of the set $\{G_{n,M} : G_{n,M} \in G(n,M)$
has property Q} will be said to be the *probability that* $G_{n,M}$ *has* Q.

If this probability tends to 1 as n → ∞, we say that *almost every* (a.e)
graph $G_{n,M}$ *has* Q.

To define the other model let p = p(n) satisfy 0 < p < 1 and
consider the set G^n of all 2^N graphs with vertex set V, where once
again V consists of n labelled vertices. Turn this set into a discrete
probability space $G(n,P(edge) = p)$ by giving a graph with m edges
probability $p^m q^{N-m}$, where q = 1-p. Thus $G(n,P(edge) = p)$ consists of
all graphs with vertex set V in which the edges are chosen independently
and with the same probability p. Vaguely speaking, $G(n,P(edge) = p)$ is
the union of all $G(n,M)$ with various weights, with most of the weight
concentrated on the models $G(n,M)$ with M ~ pN.

An element of $G(n,P(edge) = p)$ will be denoted by $G_{n,p}$. Statements
of the kind "the probability that $G_{n,p}$ has property Q is u" and "a.e.
$G_{n,p}$ has Q" need no explanation.

The case p = ½ deserves special attention for $G(n,P(edge = ½))$ is
exactly G^n with all graphs equiprobable. Thus in $G(n,P(edge) = ½)$ the
probability of Q is precisely the proportion of graphs in G^n having Q.

It is worth noting that when studying the model $G(n,M)$ we almost
always take M as a function of n, but in $G(n,P(edge) = p)$ the probability
p is often taken to be fixed. In fact, it is clear that if M is fixed
then a.e. graph $G_{n,M}$ consists of M independent edges so the model $G(n,M)$
is rather unexciting.

The models $G(n,M)$ and $G(n,P(edge) = p)$ are very close to each other
if M ~ pN. We shall not elaborate on this but will say only that in
most cases it is a simple matter to carry over an assertion from
$G(n,P(edge) = p)$ to $G(n,\lfloor pN \rfloor)$. The advantage of this is that we are
allowed to work in the model $G(n,P(edge) = p)$, where the calculations
tend to be simpler and still deduce a result about $G(n,M)$, the finer
model of random graphs.

§1: THE AUTOMORPHISM GROUP

Almost all results in probabilistic graph theory concern graphs
with labelled vertices. This would be a rather unsatisfactory state
of affairs but for the fact that results about labelled graphs can
often be translated into results about unlabelled graphs. Better
undergraduates have no difficulty in showing that almost every labelled
graph has a trivial automorphism group: for a.e. G ∈ $G(n,P(edge) = ½)$
the group Aut G consists only of the identity. Similarly, it is not
hard to show that if neither p nor q is too small then the automorphism group of a.e. G ∈ $G(n,P(edge) = p)$ is trivial. However, it
is quite remarkable that one can determine the minimal p for which
this assertion is true. This deep result is due to Wright *[77]*. To
add to the complications this is one of the cases when the model

$G(n, P(\text{edge}) = p)$ cannot be used to derive information about $G(n, \lfloor pN \rfloor)$. Denote by $U(n,M)$ the set of unlabelled graphs of order n and size M, that is the set of isomorphism classes of graphs in $G(n,M)$.

Theorem 1.1. $|U(n,M| \sim |(n,M)|/n!$ if and only if

$\min\{M, N-M\}/n - (\log n)/2 \to \infty$

as $n \to \infty$. □

It is very easy to see that the condition is necessary. Indeed, if, say, $M \leq \frac{1}{2}n \log n + cn$ for some constant c, then the probability that $G_{n,M}$ has two isolated vertices is at least some $\alpha(c) > 0$ if n is sufficiently large. (See the remark after Theorem 2.2.) As a graph with two isolated vertices has at least one non-trivial automorphism, namely the interchange of those two vertices, this shows that

$|U(n,M)| \geq (1 + \alpha(c)/2 + o(1))|G(n,M)|/n!$.

The striking fact is that this trivial necessary condition is in fact sufficient.

Let us turn $U(n,M)$ into a probability space by giving all graphs the same probability. Given a property Q of graphs, denote by $P_M^L(Q)$ the probability of Q in $G(n,M)$ and by $P_M^U(Q)$ its probability in $U(n,M)$. Theorem 1.1 has the following immediate consequence.

Corollary 1.2. Suppose

$\min\{M, N-M\}/n - (\log n)/2 \to \infty$.

Then

$P_M^U(Q) = P_M^L(Q) + o(1)$. □

This corollary enables us to use the model $G(n,M)$ instead of $U(n,M)$ unless M is too small or too large.

Recently Cameron [29] studied an interesting question related to Theorem 1.1. Given a finite group A, denote by $a_n(A)$ the proportion of those graphs $G \in G^n$ with $A \leq$ Aut G which satisfy A = Aut G. An old theorem of Frucht [48] states that for a fixed group A we have $a_n(A) > 0$ for some n and, in fact, $a_n(A) > 0$ if n is sufficiently large. The weak form of Theorem 1.1. mentioned above says that if A is the trivial group then $\lim_{n\to\infty} a_n(A) = 1$. Some of the beautiful and somewhat unexpected results of Cameron [29] are collected in the following theorem.

Theorem 1.3. (i) $a(A) = \lim_{n\to\infty} a_n(A)$ exists for every finite group A.

(ii) $a(A) = 1$ if and only if A is a direct product of symmetric groups.

(iii) If A is abelian but not an elementary abelian 2-group, then $a(A) = 0$.

(iv) The values of $a(A)$ are dense in $[0,1]$.

§2: SPARSE GRAPHS

The paper that influenced most the theory of random graphs was written by Erdős and Rényi [38] in 1960. In this paper Erdős and Rényi studied in detail the structure of a random graph $G_{n,M}$. Viewing $G_{n,M}$ as an organism that develops by acquiring edges in a random fashion, they asked at what stage of development a given property is likely to appear. Among others they showed that if $M = o(n^{\frac{1}{2}})$ then a.e. $G_{n,M}$ consists of isolated edges, if $M \sim cn^{\frac{1}{2}}$ then trees of order 3 appear with probability bounded away from 0 (see Theorem 3.1), and so on. If $M = o(n)$ then a.e. $G_{n,M}$ is a union of trees and cycles are likely to appear if $M = O(n)$. Perhaps the most surprising fact is that a certain "double jump" occurs in the size of the largest component as M passes through $\frac{1}{2}n$. More precisely, if $M \sim cn$ for some constant $0 < c < \frac{1}{2}$ then the maximal order of a component of a.e. $G_{n,M}$ is of order $\log n$, if $M \sim cn$ then the maximal size of a component has order $n^{2/3}$ and if $M \sim cn$ for some $c > \frac{1}{2}$ then a.e. $G_{n,M}$ has a component of order n. In fact in the last case a.e. $G_{n,M}$ has one giant component and the small components contain at most one cycle. Furthermore, the order of the giant component behaves rather regularly, as the following result shows.

Theorem 2.1. Denote by $|C(G)|$ the maximal order of a component of G. If $c > \frac{1}{2}$ is a constant and $M(n) \sim cn$ then for every $\varepsilon > 0$ we have
$$P(\,|\,|C(G_{n,M})| - (1 - \frac{x(c)}{2c})n| > \varepsilon n) = o(1),$$
where

$$x(c) = \sum_{k=1}^{\infty} \frac{k^{k-1}}{k!}(2ce^{-2c})^k. \qquad \square$$

Note that if c is large then $x(c)$ is close to $2ce^{-2c}$ so the number of vertices not belonging to the giant component is about $e^{-2c}n$. Even more, as c grows, eventually only the isolated vertices do not belong to the giant component. As soon as $M(n)$ is large enough to guarantee that almost no $G_{n,M}$ has an isolated vertex, a.e. $G_{n,M}$ becomes connected. In fact, Erdös and Rényi [39] proved the following precise result. As usual, $\delta(G)$ is the minimal degree and $\kappa(G)$ is the connectivity of a graph G.

Theorem 2.2. Let c be a real constant and let r be a non-negative integer. If $M(n) = \frac{1}{2}n \log n + \frac{r}{2}n \log \log n + cn + o(n)$ then
$$\lim_{n\to\infty} P(\delta(G_{n,M}) = r) = 1 - \exp(-e^{-2c}/r!)$$
and
$$\lim_{n\to\infty} P(\kappa(G_{n,M}) = r) = 1 - \exp(-e^{-2c}/r!). \qquad \square$$

This result shows that, in particular, in the range above a.e. $G_{n,M}$ has the same minimal degree and vertex connectivity. In fact, putting together results from [12] and [51], we see that this assertion holds for every function $M(n)$.

In Theorem 2.2 one can claim somewhat more, namely that the distribution of the number of vertices of degree r tends to the Poisson distribution with mean $e^{-2c}/r!$. This shows that if $M(n) \leq \frac{1}{2}n \log n + cn$

for some constant c then $\lim_{n\to\infty} P(G_{n,M}$ has at least 2 isolated vertices)

$\geq 1 - e^{-\lambda}(1+\lambda) > 0$, where $\lambda = e^{-2c}$. Hence the necessity of the condition in Theorem 1.1. does follow easily.

An immediate consequence of Theorem 2.2. is that in order to guarantee that a.e. $G_{n,M}$ has minimal degree at least r the function $M(n)$ has to satisfy $M(n) \geq \frac{n}{2}\{\log n + (r-1) \log \log n + \omega(n)\}$, where $\omega(n) \to \infty$. Erdős and Rényi [41] proved that for r = 1 the same condition guarantees also that a.e. $G_{n,M}$ has a 1-factor. Recently this result was extended considerably by Shamir and Upfal [72].

Theorem 2.3. Let r ∈ IN be fixed,

$$M(n) \geq \frac{n}{2} \log n + (r-1) \log \log n + \omega(n) ,$$

where $\omega(n) \to \infty$, and let $f_n : \{1,2,\ldots,n\} \to \{1,\ldots,r\}$ be such that

$\sum_{i=1}^{n} f_n(i)$ is even. Then a.e. $G_{n,M}$ contains an f-factor, i.e. a subgraph in which vertex i ∈ V = $\{1,\ldots,n\}$ has degree $f_n(i)$ □

One of the hard problems left open by Erdős and Rényi concerned the giant component in Theorem 2.2. Does this component contain a long path? This question has been solved only recently; the following remarkable result was proved by Ajtai, Komlós and Szemerédi, and, independently, by de la Véga [76].

Theorem 2.4. There is a function $\alpha = \mathbb{R}^+ \to \mathbb{R}$ with $\lim_{c\to\infty} \alpha(c) = 1$ such that if $M(n) \geq cn$ for some constant c then a.e. $G_{n,M}$ contains a path of length $\lfloor \alpha(c)n \rfloor$.   □

Corollary 2.5. If $M(n)/n \to \infty$ then a.e. $G_{n,M}$ contains a path of length $(1+o(1))n$.

Let us conclude this section with a beautiful recent result of Erdős and Spencer showing another "double jump" similar to the one mentioned at the beginning of the section. Let C_n be the graph of the n-dimensional cube. Thus the vertices of C_n are sequences $(\varepsilon_1, \varepsilon_2, \ldots, \varepsilon_n)$ with $\varepsilon_i = 0$ or 1, with two sequences adjacent iff they differ in exactly one term. Given a constant p, $0 < p < 1$, let $C_{n,p}$ be a random graph in which an edge of C_n is present with probability p, independently of the other edges, and which contains no edge not belonging to C_n. Extending results of Burtin [28], Erdős and Spencer [44] proved that the probability of $C_{n,p}$ being connected is a remarkable function of p.

Theorem 2.6.

$$\lim_{n\to\infty} P(C_{n,p} \text{ is connected}) = \begin{cases} 0 & \text{if } p < \tfrac{1}{2}, \\ 1/e & \text{if } p = \tfrac{1}{2}, \\ 1 & \text{if } p > \tfrac{1}{2}. \end{cases}$$

    □

§3: THRESHOLD FUNCTIONS

A property Q of graphs of order n is usually identified with the subset of G^n consisting of the graphs that possess Q. We call Q *monotone increasing* if $G \in Q$, $H \in G^n$ and $G \subset H$ imply that $H \in Q$. For example, the properties of being connected or containing a given graph or having diameter at most d are all monotone increasing. Given a monotone increasing property Q and a function A(n) > 0, Erdős and Renyi [38] call A(n) a *threshold function* for Q if

$$\lim_{n \to \infty} P(G_{n,M} \text{ has } Q) = \begin{cases} 0 \text{ if } M(n)/A(n) \to 0, \\ 1 \text{ if } M(n)/A(n) \to \infty. \end{cases}$$

Ideally one would like to determine considerably more than a threshold function.

Let F(c) be a continuous distribution function and let $F = \{M_c : 0 < F(c) < 1\}$ be a family of integer valued functions defined on the natural numbers. We call the pair (F,F) an *exact probability distribution* for the property Q if

$$\lim_{n \to \infty} P(G_{n,M_c} \text{ has } Q) = F(c).$$

for every c with 0 < F(c) < 1. Note that $M_c(n)$ need not be positive for every n but it has to be positive if n is sufficiently large. Furthermore, we call $M_c(n)$ an *exact threshold function with probability* F(c).

Though an exact probability distribution is clearly not unique, it tells us almost everything we wish to know about the property. For example, if M(n) is any integer valued function,

$$\overline{\lim_{n \to \infty}} P(G_{n,M} \text{ has } Q) = F(\beta),$$

where

$$\beta = \sup\{c : \overline{\lim_{n \to \infty}} (M(n)) \geq 0\}.$$

For example, if Q is the property of being connected then Theorem 2.2. shows that $M_n(n) = \frac{n}{2}(\log n + c)$ is an exact threshold function with probability $F(c) = \exp(-e^{-c})$.

In this section we shall discuss the existence of threshold functions for being Hamiltonian, containing a given graph and having diameter at most d.

Let us begin with the property of containing a graph isomorphic to a given graph. We define the (average) *degree* of a graph H as d(H) = 2e(H)/|H|. The *maximal subgraph degree of* H is m(H) = max{d(F) : F ⊂ H}. A graph H is said to be *strictly balanced* if d(H) > d(F) for every F ⊂ H, F ≠ H. Clearly trees, cycles and complete subgraphs are all strictly balanced. Let Q_H be the property that our graph contains a subgraph isomorphic to H. Erdős and Rényi [38] determined a threshold function for Q_H, whenever H is a strictly balanced graph. Furthermore, Schürger [71] determined an exact threshold function for Q_H in the case H is a complete graph. The following theorem from [18] extends these results.

Theorem 3.1. Let H be a fixed graph with a unique subgraph of maximal degree. Suppose H has k vertices and $\ell \geq 2$ edges, and the automorphism group of H has a elements. Then $M_c(n) = \lfloor cn^{2-k/\ell} \rfloor$, $c > 0$ is an exact threshold function for Q_H with probability $1 - \exp(-(2c)^\ell/a)$:
$$\lim_{n \to \infty} P(G_{n,M_c} \text{ has } Q_H) = 1 - \exp(-(2c)^\ell/a). \qquad \Box$$

Considerably more is true than we have just stated, namely the number of subgraphs isomorphic to H tends to the Poisson distribution with mean $(2c)^\ell/a$. In fact, the natural proof gives exactly this result. One shows that for every fixed $r \geq 0$ the rth factorial moment $E_r(X(X-1)...(X-r+1))$ of $X = X(G)$, the number of subgraphs of G isomorphic to H, tends to λ^r as $n \to \infty$. Hence X tends in distribution to the appropriate Poisson distribution.

The simplest case of the theorem above is when H is a tree of order 3. Then $k = 3, \ell = 2$ and $a = 2$ so if $M_c(n) = \lfloor cn^{\frac{1}{2}} \rfloor$, $c > 0$, then the probability that a graph G_{n,M_c} contains a tree of order 3 tends to e^{-2c^2}. In particular, as $n \to \infty$, the probability is bounded away from 0, as noted at the beginning of §2. If H is a cycle of order k then the probability that a graph $G_{n,\lfloor cn \rfloor}$ contains H tends to $1 - \exp\{-(2c)^k)/(2k)\}$. Let us note one more immediate consequence of Theorem 3.1. If H is any graph containing two cycles with at least one vertex in common and
$$\lim_{n \to \infty} (G_{n,M} \text{ has } Q_H) > 0$$
then for some $\varepsilon > 0$ we have $Mn^{-1-\varepsilon} \to \infty$

For an arbitrary graph H one has the following weaker result.
Theorem 3.2. Let H be an arbitrary graph with $d(H) = 2\ell/k > 1$. Then the probability of containing H has threshold function $n^{2-k/\ell}$. $\qquad \Box$

What is the threshold function of a Hamilton cycle? This problem has attracted a lot of attention, resisting attempts for many years. After several partial results due, among others, to Moon [64], Wright [78] and Perepelica [65], the back of the problem was broken by Pósa [67] and Korshunov [57].

Let us note first that if a.e. $G_{n,M}$ contains a Hamilton cycle then, a fortiori, a.e. $G_{n,M}$ has minimal degree at least 2. Hence, by Theorem 2.2, we must have $M \geq \frac{n}{2}\{\log n + \log n + \omega(n)\}$, where $\omega(n) \to \infty$. Pósa and Korshunov proved that this rather trivial necessary condition is almost sufficient.

Theorem 3.3. If $M \geq 10n \log n$ then a.e. $G_{n,M}$ contains a Hamilton cycle. $\qquad \Box$

Here 10 is just a constant one can obtain without much trouble, it is not the best possible value the proofs give. Subsequently Angluin and Valiant [4] found a fast probabilistic algorithm for a Hamilton cycle. The point of this algorithm is that it looks for a Hamilton cycle by testing every edge only once and with probability tending to 1 it finds a Hamilton cycle in a $G_{n,M}$ graph with $M \geq cn$ log n. Other algorithms in this vein have been produced by Shamir and de la Véga. Recently Theorem 3.3 has been sharpened by Korshunov [58] and Komlós and Szemerédi. In particular, they proved that if $M \geq \frac{n}{2}\{\log n + \log \log n + \omega(n)\}$ for some $\omega(n) \to \infty$, then a.e. $G_{n,M}$ contains a Hamilton cycle.

The algorithm given by Angluin and Valiant proves also that almost every *directed* graph with cnlog n edges has a Hamilton cycle. Very recently McDiarmid [62], [63] proved some general theorems with bearing on the problem of Hamilton cycles in random directed graphs.

Given $d \geq 2$, for what function M = M(n) is it true that a.e. $G_{n,M}$ has diameter at most d? This question was studied by Moon and Moser [65], Korshunov [56] and, very recently, by Klee and Larman [55]. The results of this last paper were considerably extended in [16], where the exact probability distribution of diam $G_{n,M}$ was determined for a large class of functions M(n). Here we state only a special case of a result giving the range of M in which a.e. $G_{n,M}$ has a given diameter.

Theorem 3.4. Let $d \geq 3$ be fixed and suppose M = M(n) satisfies

$$M^d n^{-d-1} - 2^{-d+1} \log n \to \infty \text{ and } M^{d-1} n^{-d} - 2^{-d+2} \log n \to -\infty.$$

Then a.e. $G_{n,M}$ has diameter d. □

The result above is best possible.

The first major application of random graphs was based on the fact that for an appropriate function M(n) most $G_{n,M}$ contain relatively few short cycles while their chromatic number is large. This enabled Erdős [34] to prove that for every $k \in \mathbf{N}$ there is a k-chromatic graph of girth at least k. Various extensions of this result can be found in Bollobás and Sauer [23] and Bollobás and Thomason [24].

§4: GRAPHS WITH MANY EDGES

In this section we shall discuss some properties of $G_{n,p}$ for p *fixed* , 0 < p < 1. The graph invariant of $G_{n,p}$ that has attracted the most attention is perhaps the *clique number*, the maximal order of a complete subgraph. The clique number of G is denoted by cl(G). Matula [59] was the first to show that the clique number has a strong peak: most $G_{n,p}$ have about the same clique number. Later Grimmett and McDiarmid [49] and Bollobás and Erdős [22] proved rather sharp results about this peak. Here we give only a weak form of a result from [22] showing that for most graphs the clique number takes only one of two possible values and, even more, for most values of n most $G_{n,p}$ have the same clique number.

Theorem 4.1. Let $d = d(n)$ be the positive real number for which

$$\binom{n}{d}p^{\binom{d}{2}} = 1.$$

Then for every $\varepsilon > 0$ a.e. $G_{n,p}$ has clique number at least $d(n)-1-\varepsilon$ and at most $d(n)+\varepsilon$. □

The rough value of $d(n)$ is easily determined:

$$d(n) = 2\log_b n - 2\log_b \log_b n + 2\log_b(\tfrac{1}{2}e) + o(1)$$

$$= 2\log_b n + O(\log\log n) = \frac{2\log n}{\log(1/p)} + O(\log\log n),$$

where $b = 1/p$. Thus Theorem 4.1 has the following consequence.

Corollary 4.2. A.e. $G_{n,p}$ satisfies

$$cl(G_{n,p}) = \frac{2\log n}{\log(1/p)} + O(\log\log n). \qquad □$$

Since the clique number of the complement (the independence number) and the chromatic number clearly satisfy $cl(\overline{G})\chi(G) \geq |G|$ for every graph G, the corollary above gives a lower bound for the chromatic number of a random graph. In both *[22]* and *[49]* the simplest colouring algorithm, the *greedy algorithm* (see *[11, p.89]*) was used to obtain an upper bound for the chromatic number of most $G_{n,p}$. Once again, we state the result in a rather weak form.

Theorem 4.3. The chromatic number of a.e. $G_{n,p}$ satisfies

$$\frac{n\log(1/p)}{2\log n} \leq \chi(G_{n,p}) \leq \frac{n\log(1/p)}{\log n}(1 + o(1)). \qquad □$$

The gap between the two sides of this inequality is most fascinating. I believe that the left-hand side gives the correct order of magnitude, but I would consider it most interesting if one could prove that for some $\varepsilon > 0$ a.e. $G_{n,p}$ has chromatic number at most $(1-\varepsilon) n \log(1/p)/\log n$. It will certainly not be easy to find a substantially more efficient algorithm than the greedy algorithm (see McDiarmid *[60]*, *[61]*).

The bounds on the chromatic number of a random graph have had some interesting applications. A conjecture attributed to Hajós stated that a k-chromatic graph contains a TK^k, a topological complete graph of order k. This conjecture was refuted by Catlin *[30]* who gave counterexamples for $k \geq 7$. A little later it was shown by Erdös and Fajtlowicz *[35]* that in fact a.e. $G_{n,\frac{1}{2}}$ is a counterexample to the conjecture since almost no $G_{n,\frac{1}{2}}$ contains a TK^k for $k \geq cn^{\frac{1}{2}}$, while by Theorem 4.3 almost every $G_{n,\frac{1}{2}}$ has chromatic number at least $n \log 2/(2 \log n)$. Subsequently Bollobás and Catlin *[20]* proved that for a.e. $G_{n,\frac{1}{2}}$ the maximal k with $TK^k \subset G_{n,\frac{1}{2}}$ is asymptotic to $2n^{\frac{1}{2}}$. In view of this it is interesting to note that Hadwiger's conjecture *is* true for almost every graph, as shown by Bollobás, Catlin and Erdős [21].

The degree sequence $d_1 \geq d_2 \geq \ldots \geq d_n$ of a random graph is fairly close to a sequence of independent binomial random variables so it is to be expected that we have detailed information about it. Erdős and Wilson [45] showed that a.e. $G_{n,p}$ has a unique maximal degree, that is $d_1 - d_2 \geq 1$ for a.e. graph. Cornuejols [32] and Bollobás [12] investigated the size of a general term and the jumps $d_1 - d_2$, $d_2 - d_3$, ... It turns out that the maximal degree of a.e. graph belongs to a rather small interval. Even more, the exact probability distribution of the maximal degree is determined in [14].

Theorem 4.4. Let c be a fixed real number. Then the probability that the maximal degree of $G_{n,p}$ is less than

$$pn + (2pqn \log n)^{\frac{1}{2}}\{1 - \frac{\log \log n}{4\log n} - \frac{\log(2\pi^2)}{2\log n} + \frac{y}{2\log n}\}$$

tends to $e^{-e^{-y}}$.

In particular, if $\omega(n) \to \infty$ then the maximal degree d_1 of a.e. $G_{n,p}$ satisfies

$$|pn + (2pqn \log n)^{\frac{1}{2}}\{1 - \frac{\log \log n}{4\log n}\} - d_1| < \omega(n)(\frac{n}{\log n})^{\frac{1}{2}}. \qquad \square$$

It may seem somewhat surprising that not only the maximal degree is unique in a.e. $G_{n,p}$, but the degree sequence starts with many strictly decreasing terms. We state only a simple corollary from [12].

Theorem 4.5. A.e. graph $G_{n,p}$ is such that $d_1 > d_2 > \ldots > d_m$ for $m = o(n^{\frac{1}{4}})/(\log n)^{\frac{1}{2}}$.

Jumps and repeated values in the degree sequence are discussed in [19].

We conclude the section with a fascinating result of Babai, Erdős and Selkow [5] (See also Babai and Kučera [6]), based on a weaker form of Theorem 4.5, discovered independently of [12].

Theorem 4.6. There is an algorithm which, for almost all graph G, tests any graph for isomorphism to G within linear time. $\qquad \square$

The expression "almost all" appearing in the theorem refers to the model $G(n,\frac{1}{2})$. The gist of the proof is that, as Theorem 4.5 shows, the degrees of many vertices are unique so we can identify many vertices by their degrees. Furthermore, almost every graph is such that no two vertices are joined to the same set of vertices of high degree, so the relationship to a few of the vertices of high degree identifies every vertex.

The model $G(n,\frac{1}{2})$ is the basis of one of the first important applications of random graphs: the lower bounds for Ramsey numbers given by Erdős [33]. For sharper lower bounds for various Ramsey numbers, all based on the probabilistic method, see Spencer [73], [74], Ajtai, Komlós and Szemerédi [1] and Burr, Erdős, Faudree,

Rousseau and Schelp [27].

§5. RANDOM REGULAR GRAPHS

Denote by $G(n,r\text{-reg})$ the set of all r-regular graphs with a fixed
set V of n vertices. We shall always assume that $r \geq 2$ and rn is even.
Thus if r is odd then $n \to \infty$ means that n tends to infinity through
even integers. The set $G(n,r\text{-reg})$ is turned into a probability space
by giving every graph $G \in G(n,r\text{-reg})$ the same probability. This
probability space consists of *random* (labelled) *r-regular graphs with
n vertices*. Whenever we ask questions about random regular graphs,
we have this model in mind. What is the expected number of Hamilton
cycles in 5-regular graphs? What is the probability that an 8-regular
graph contains a triangle? We know now that these questions are
quite easy, but until recently they could not be tackled because no
amenable representation of the space $G(n,r\text{-reg})$ was available. Even
more, until two years ago there was no asymptotic formula for the
number of (labelled) regular graphs.

This situation was rather curious since over twenty years ago
Read [68] did determine an exact formula for $|G(n,r\text{-reg})|$, the number
of labelled r-regular graphs of order n. This formula, whose proof
is based on Pólya's enumeration theorem [66], is rather complicated
and it does not lend itself readily to asymptotics. Bender and
Canfield [8] were the first to give an asymptotic formula for the
number of labelled graphs with given degree sequences, so in particular
for the number of labelled r-regular graphs. The ingenious proof in
[8] is based on enumerating certain classes of involutions and it is
not clear that it can be used to tackle questions concerning random
regular graphs. Recently in [15] a probabilistic proof was given for
the asymptotic number of labelled regular graphs. This proof is based
on a rather natural representation of $G(n,r\text{-reg})$ as the image of a
considerably simpler set. This approach enables one to obtain a more
general asymptotic formula without much effort and without any reference
to exact formulae. In particular, the formula one obtains holds not
only for constant values of r but also if $r = r(n)$ is a function of n
bounded by $(2 \log n)^{\frac{1}{2}} - 1$. So far we have mentioned only regular
graphs but it is worth noting that the same method is applicable to
graphs with a given degree sequence. However, as the calculations
are less attractive for a general degree sequence, we concentrate on
regular. graphs. Furthermore, we shall keep r fixed for that simplifies
the calculations considerably.

The representation of $G(n,r\text{-reg})$ is very simple indeed. We shall
obtain the graphs in $G(n,r\text{-reg})$ as images of so-called configurations.
Let $W = \bigcup_{j=1}^{n} W_j$ be a fixed set of rn labelled vertices, where $|W_j| = r$.

A *configuration* F is a partition of W into $m = rn/2$ pairs of vertices, called *edges* of F. Let $\Phi = \Phi(n,r)$ be the set of all configurations and let $\Omega = \Omega(n,r)$ be the set of those configurations which do not contain an edge joining two vertices of some W_j or a pair of edges joining vertices in the union of two sets W_j. This set Ω is very closely related to $G(n,r\text{-reg})$. Given $F \in \Omega$, define a graph $\phi(F)$ with vertex set $V = \{W_1, W_2, \ldots, W_n\}$ in which W_i is joined to W_j if F contains an edge joining a vertex of W_i to a vertex of W_j. Then $\{\phi(F) : F \in \Omega\}$ is exactly $G(n,r\text{-reg})$, that is *the set of all r-regular graphs with vertex set* V.

What have we gained by constructing $G(n,r\text{-reg})$ in this manner? At the first sight we have complicated rather than simplified our model since though ϕ is simple enough, from Φ we have to pass to Ω and then to $\phi(\Omega)$. However, it is immediate that each graph $G \in G(n,r\text{-reg})$ comes from $|\phi^{-1}(G)| = (r!)^n$ configurations. Furthermore, and this is the key to this view of $G(n,r\text{-reg})$,

$$|\Omega| \sim e^{-(r^2-1)/4}|\Phi|. \tag{1}$$

Since clearly

$$|\Phi| = (2m)_m 2^{-m}, \tag{2}$$

where as before $m = rn/2$, (1) implies that

$$|G(n,r\text{-reg})| \sim e^{-(r^2-1)/4}(2m)!/\{m!(r!)^n 2^m\}. \tag{3}$$

Before proving relation (1) we emphasize that the point of this approach is not so much the proof of (3) but the possibility of using the simpler model $\Phi(n,r)$ in studying questions about $G(n,r\text{-reg})$. Even more, if r is fixed then in most questions instead of (1) it suffices to use the fact that in Φ considered as a probability space the probability that a configuration F is mapped into a graph by ϕ (rather than into a graph with loops and multiple edges) is bounded away from 0. Thus for example this rather weak property implies that if almost every configuration has $\lfloor n/2 \rfloor$ independent edges (defined in the natural way) then almost every r-regular graph has $\lfloor n/2 \rfloor$ independent edges.

Given a natural number $\ell \geq 3$ and a graph G, denote by $X_\ell(G)$ the number of cycles of length ℓ in G. Note that $X_3, X_4, \ldots$ are random variables on $G(n,r\text{-reg})$. Our main aim in this section is to show (1) and to determine the asymptotic joint distribution of these random variables $X_3, X_4, \ldots$. We shall see that a slight extension of the proof of (3) tells us this asymptotic joint distribution. As we take r fixed, the proof below follows that in *[13]* rather than in *[15]*.

Theorem 5.1. Let r be a fixed natural number and let $n \to \infty$ such

that $m = m(n) = rn/2$ is also an integer. Then the number of labelled
r-regular graphs of order n is

$|G(n,r\text{-reg})| \sim e^{-(r^2-1)/4}(2m)!/\{m!(r!)^n 2^m\}.$

Furthermore, for any fixed k the random variables $X_3, X_4, \ldots, X_k$ tend in
distribution to k-2 independent Poisson random variables $X_3^*, X_4^*, \ldots, X_k^*$
with X_ℓ^* having mean $(r-1)^\ell/2\ell$.

Proof. Note that there are

$N = N(m) = (2m)_m 2^{-m}$

configurations and the number of configurations containing t given
independent edges (i.e. t given disjoint pairs of vertices) is

$N(m-t) = (2m-2t)_{m-t} 2^{-m+t}.$

Hence for a fixed t

$P_t = N(m-t)/N(m) = 2^t(m)_t/(2m)_{2t} \sim (rn)^{-t}$

is the probability that a configuration contains t given independent
edges.

An *ℓ-cycle of a configuration* F is a set of ℓ edges of F say
$e_1, e_2, \ldots e_\ell$, such that for some ℓ distinct groups $W_{j_1}, W_{j_2}, \ldots W_{j_\ell}$ the
edge e_i joins W_{j_i} to $W_{j_{i+1}}$, where $W_{j_{\ell+1}}$ is defined to be W_{j_1}. Note
that this definition makes sense for every natural number ℓ. Denote
by $Y_\ell(F)$ the number of ℓ-cycles of a configuration F. It is clear that
$\Omega = \{F \in \Phi : Y_1(F) = Y_2(F) = 0\}.$

that is $\phi(F)$ is a graph rather than a multigraph if and only if F has
no 1-cycles and 2-cycles. Furthermore, if $F \in \Omega$ then $Y_\ell(F) = X_\ell(\phi(F))$
for every $\ell \geq 3$. Therefore both assertions of the theorem follow if we
show that the random variables $Y_1, Y_2, \ldots, Y_k$ (defined on Φ) tend in
distribution to k independent Poisson random variables $Y_1^*, Y_2^*, \ldots, Y_k^*$
with Y_ℓ^* have mean $\lambda_\ell = (r-1)^\ell/2\ell$. According to the Jordan inequalities
[52],[53],[54] (see also Bonferroni [25] and Chung [31,p.99]), this
holds if and only if the joint moments of $Y_1, Y_2, \ldots, Y_k$ tend to the
joint moments of the appropriate Poisson variables. Thus if
$E(s_1, s_2, \ldots, s_k)$ denotes the expected number of sets consisting of s_1
1-cycles, s_2 2-cycles,..., s_k k-cycles, then we have to show that

$$E(s_1, s_2, \ldots s_k) \sim \prod_{\ell=1}^{k} \{\lambda_\ell^{s_\ell}/s_\ell!\} \qquad (4)$$

whenever $s_1, s_2, \ldots s_k$ are fixed non-negative integers.

Let us write

$$E(s_1, s_2, \ldots, s_k) = E_1 + E_2, \qquad (5)$$

where E_1 is the expected number of sets consisting of s_1 1-cycles,
s_2 2-cycles,..., s_k k-cycles such that no group W_i contains vertices

of two cycles. We have

$$((n)_\ell/2\ell)(r(r-1))^\ell$$

choices for an ℓ-cycle so

$$(n)_t(r(r-1))^t \prod_{\ell=1}^{k}(2\ell)^{-s_\ell} \prod_{\ell=1}^{k}(s_\ell!)^{-1}$$

choices for the sets counted in E_1, where

$$t = \sum_{\ell=1}^{k} \ell s_\ell$$

is the number of groups W_j containing vertices of these cycles and also the total number of edges in these cycles. Hence

$$E_1 = (n)_t\ (r(r-1))^t \prod_{\ell=1}^{k}(2\ell)^{-s_\ell} \prod_{\ell=1}^{k}(s_\ell!)^{-1}P_t$$

$$\sim\ (rn)^t \prod_{\ell=1}^{k}((r-1)^\ell/2\ell)^{s_\ell} \prod_{\ell=1}^{k}(s_\ell!)^{-1}P_t$$

$$\sim\ \prod_{\ell=1}^{k}\{(\frac{(r-1)^\ell}{2})^{s_\ell}/s_\ell!\}\quad = \prod_{\ell=1}^{k}\{\lambda_\ell^{s_\ell}/s_\ell!\}. \tag{6}$$

Relations (5) and (6) show that (4) holds if and only if $E_2 = o(1)$. To see this note that a set of cycles counted in E_2 is such that the vertices of the cycles are contained in some $u \le t-1$ sets W_j and the cycles have at least $u+1$ edges. We use this to obtain a very crude estimate for E_2. Given a set of u groups W_j there are at most

$$(ru)^{2t}$$

sets of cycles contained in these groups. Hence

$$E_2 \le \sum_{u=1}^{t-1} \binom{n}{u}(ru)^{2t}P_{u+1} = \sum_{u=1}^{t-1} O(n^{-1}) = O(n^{-1}) = o(1),$$

completing the proof of (4).

Though we have already remarked that the assertions of the theorem follow from (4), we spell out the deduction of the first assertion in detail.

$$P_\phi(\Omega) = P_\phi(Y_1 = Y_2 = 0) \to e^{-\lambda_1}e^{-\lambda_2} = e^{-(r^2-1)/4}$$

so

$$|\Omega| \sim e^{-(r^2-1)/4}|\phi|.$$

Every r-regular graph G with vertex set $\{W_1,W_2,\ldots,W_n\}$ is the image of some configuration $F \in \Omega$. Furthermore, if π is a permutation of W leaving each W_j invariant, then π maps F into a configuration $\pi(F)$ satisfying $G = \phi(F) = \phi(\pi(F))$ and $\pi(F) = F$ implies that π is the identity. Since we have $(r!)^n$ choices for π, $|\phi^{-1}(G)| = (r!)^n$. This shows (2), concluding the proof of the theorem.

As an example of the use of the model $\Omega(n,r)$ we prove an intuitively obvious fact.

Theorem 5.2. Let $r \geq 3$ be fixed and let $n \to \infty$ in such a way that rn is even for every n. Then

$$P(G_{n,r\text{-}reg} \text{ is connected}) \to 1.$$

Proof. If $n \geq r+1$ and $G_{n,r\text{-}reg}$ is not r-connected then it comes from a configuration $F \in \Omega(n,r)$ for which $V = \{W_1 W_2, \ldots, W_n\}$ can be partitioned into non-empty sets V_0, V_1, and V_2 such that $|V_0| = r-1$, $|V_1| = m_1 \leq |V_2| = m_2 = n - m_1 - r + 1$ and whenever $W_i \in V_1$ and $W_j \in V_2$, F contains no $W_i - W_j$ edge. Furthermore, as every vertex of $G_{n,r\text{-}reg}$ has degree r, we must have $m_1 > 2$. We shall show that the probability of such a configuration tends to 0.

In order to avoid some inessential complications, we shall assume that r is even. Denote by $R(m_1)$ the number of configurations that do contain V_0, V_1 and V_2 as above, with $|V_1| = m_1 \leq |V_2| = m_2 = n - m_1 - r + 1$. Furthermore, denote by $R(m_1, \ell_1, \ell_2)$ the number of these with ℓ_1 edges from V_0 to V_1 and ℓ_2 edges from V_0 to V_2. Thus

$$R(m_1) = \sum_{\ell_1, \ell_2} R(m_1, \ell_1, \ell_2),$$

where the summation is over all ℓ_1 and ℓ_2 with $\ell_1 + \ell_2 \leq r(r-1)$.

For a fixed $t \geq 1$ the probability that some t groups W_i span at least $t+1$ edges is at most

$$\binom{n}{t}(rt)^{2(t+1)}N(rn-2(t+1))/N(rn) = O(n^{-1}).$$

This implies that for every fixed $m_1 \geq 2$

$$R(m_1)/N(rn) = O(n^{-1})$$

since every configuration counted in $R(m_1)$ contains a set of $t \leq m_1 + r - 1$ groups spanning at least $t+1$ edges. Consequently

$$\sum_{m_1=2}^{r^3} R(m_1) = o(N(rn)).$$

There remains to show that

$$\sum_{m_1=r^3}^{\lfloor (n-r+1)/2 \rfloor} R(m_1) = o(N(rn)).$$

We start with the crude bound

$$R(m_1, \ell_1, \ell_2) \leq \binom{n}{r-1}\binom{n}{m_1}(r(r-1))!(rm_1)^{\ell_1}(rm_2)^{\ell_2}N(rm_1-\ell_1)N(rm_2-\ell_2).$$

Easy calculations give

$$R(m_1) \leq K_1 n^{r-1}\binom{n}{m}n^{r(r-1)}N(rm_1)N(rm_2-r(r-1)) = S(m_1),$$

where K_1 depends only on r. Now

$$S(m_1+1)/S(m_1) = \frac{n-m_1}{m_1+1} \frac{rm_1+r-1}{rn-rm_1-r+1}$$

and this implies that in the range $r^3 \leq m_1 \leq n/2$ we have

$\quad S(m_1) \leq \max\{S(r^3), S(\lfloor n/2 \rfloor)\}$.

It is easily checked that the right-hand side is $o(N(rn)/n)$. Hence
(7) follows, completing the proof.

Corollary 5.3. Let $r \geq 3$ be fixed, $\ell \to \infty$ and $n = 2\ell$. Then almost
every $G_{n,r\text{-reg}}$ has a 1-factor.

Proof. It is well known (see [9, p.88]) that an r-connected r-regular
graph of order n has $\lfloor n/2 \rfloor$ independent edges, so the assertion follows
from Theorem 5.2.

It is likely that Corollary 3 can be strengthened considerably.

Conjecture. For a fixed $r \geq 3$ almost every $G_{n,r\text{-reg}}$ is Hamiltonian.

The evidence for this conjecture is rather substantial but not
overwhelming. In particular, as we shall remark later, in $G(n,r\text{-reg})$
the expected number of Hamilton cycles tends to infinity and the
variance is about a constant times the square of the expectation.
Unfortunately the constant is positive for every $r \geq 3$.

The method of proof of Theorem 5.1 is easily applied to a number
of problems concerning regular structures. For example, denote by
$C(n,r)$ the number of r-regular graphs whose edges are (properly)
coloured with r distinguishable colours. We emphasize that $C(n,r)$
is the number of *r-edge-coloured* graphs and not the number of r-edge-
colourable graphs. Then, as remarked in [13], we have the following
result.

Theorem 5.4. If $r \geq 1$ is fixed and n is even, then

$\quad C(n,r) \sim e^{(1-r)r/4} \, (n!/\{(n/2)! \; 2^{n/2}\})^r$.

Proof. Consider the set M of (properly) r-coloured r-regular multigraphs on $\{1,2,\ldots,n\}$.

Since the edges of the same colour form a 1-factor, and there are
$n!/\{(n/2)! \; 2^{n/2}\}$ possible 1-factors, we have

$\quad |M| \sim (n!/\{(n/2)! \; 2^{n/2}\})^r$.

We are interested in the number of graphs in M, that is in the
number of elements of M that do not contain multiple edges. In order
to do this we view M as a probability space and denote by $Z(M)$ the
number of *pairs of edges* of $M \in M$ joining the same two vertices. It
is easily checked that the random variable M tends in distribution to
the Poisson random variable with mean $r(r-1)/4$. We shall not check this
in detail, but show only that the expectation of M tends to $r(r-1)/4$..
There are

$\quad \binom{n}{2}\binom{r}{2}$

choices for a pair of *coloured* edges joining the same two vertices. The
choice of these two edges does not affect the selection of the edges
of different colours, while the edges of one of these colours can be

chosen in
$$(n-2)!/\{((n-2)/2)! \ 2^{(n-2)/2}\}$$
ways. Hence
$$E(M) = \binom{n}{2}\binom{r}{2} \ \{\frac{(n-2)!}{((n-2)/2)! \ 2^{(n-2)/2}} \ \frac{(n/2)! \ 2^{n/2}}{n!}\} \sim \frac{r(r-1)}{4} \ ,$$
as claimed.

The fact that M tends to the Poisson distribution with mean $r(r-1)/4$ implies that
$$C(n,r) = P(M = 0)|M| \sim e^{(1-r)r/4}(n!/\{(n/2)! \ 2^{n/2}\})^r.$$
These two results and Stirling's formula have the following immediate consequence.

Corollary 5.5. For a fixed r the expected number of r-edge-colourings of an r-regular graph of even order n is asymptotic to
$$e^{-(r-1)/4} \ 2^{(r-1)/2} \ (r!/r^{r/2})^n.$$

A. Békéssy, P. Békéssy and Komlós [7] determined the asymptotic number of regular bipartite graphs. For the sake of simplicity, we consider the set of r-regular n by n bipartite graphs with fixed vertex classes. Let $U_1, U_2, \ldots, U_n$, $V_1, V_2, \ldots, V_n$ be disjoint groups of r vertices each. Consider the set Ψ of 1-1 maps of
$$\overset{n}{\underset{1}{\cap}}U_i \text{ onto } \overset{n}{\underset{1}{\cap}}V_i.$$
Clearly $|\Psi| = (rn)!$ If for some $\psi \in \Psi$ no two vertices belonging to the same group U_i are mapped into the same group V_j then ψ determines an n by n r-regular bipartite graph with vertex classes $\{U_1, U_2, \ldots, U_n\}$ and $\{V_1, V_2, \ldots, V_n\}$: U_i is joined to V_j if $\psi(U_i) \cap V_j \neq \emptyset$. Furthermore, every r-regular bipartite graph with these vertex classes comes from exactly $r!^{2n}$, such maps in ψ. For $\psi \in \Psi$ denote by $Z(\psi)$ the number of pairs of vertices (x,y) with $x,y \in U_i$ and $\psi(x),\psi(y) \in V_j$ for some i and j. It is easily seen that, with ψ viewed as a probability space,
$$E(Z) = n^2\binom{r}{2}^2 2/(rn)(rn-1) \sim \frac{(r-1)^2}{2}$$
Furthermore, Z tends to the Poisson distribution with mean $(r-1)^2/2$. Hence follows the result of Békéssy, Békéssy and Komlós: the number of r-regular labelled n by n bipartite graphs is
$$P(Z = 0)|\psi| \sim e^{-(r-1)^2/2} \ (rn)!/r!^{2n}.$$

Exactly the same model was used by Schrijver and Valiant [70] to show that there exists an n×n matrix with non-negative integer coefficients such that each row and column sum equals r and the permanent is at most
$$r^{2n}/\binom{rn}{n}.$$

In fact, in a work under progress, McKay and I proved that for a fixed r there is an n×n 0-1 matrix with r 1's in each row and column and permanent at most

$$(1 + o(1))\ e^{-\frac{1}{2}}\ r^{2n}/\binom{rn}{n}.$$

This follows from the fact that for a fixed r the expected number of 1-factors in an r-regular n by n bipartite graph is asymptotic to
$$e^{-\frac{1}{2}}\ r^{2n}/\binom{rn}{n}\ .$$

(If one considers multigraphs instead of graphs then the factor $e^{-\frac{1}{2}}$ is replaced by 1 - hence the result of Schrijver and Valiant.)

A related result, due to Schrijver [69], concerns the minimal number of r-edge-colourings of an r-regular n by n bipartite multigraph. By determining the expected number of colourings Schrijver showed that this minimum is at most

$$r!^{2n}\ n!^r/(rn)!\ .$$

In fact, for a fixed r the expected number of r-edge-colourings of an r-regular n by n bipartite *graph* is asymptotic to
$$e^{-(r-1)/2}\ r!^{2n}\ n!^r/(rn)!\ .$$

McKay and I have calculated the expectation and variance of a number of natural random variables on $G(n,r\text{-reg})$. Write X_H for the number of Hamilton cycles, X_F for the number of 1-factors and X_α for the number of sets of $\lfloor\alpha n/2\rfloor$ independent edges. Then

$$E(X_h) \sim e(\frac{\pi}{2n})^{\frac{1}{2}}\ (\frac{(r-1)^{r-2}(r-1)^2}{r^{r-2}})^{n/2}\ ,$$

if n is even then

$$E(X_F) \sim e^{\frac{1}{4}}\sqrt{2}\ \ (\frac{(r-1)^{r-1}}{r^{r-2}})^{n/2}$$

and if $0 < \alpha < 1$ is fixed then

$$E(X_\alpha) \sim \exp\{\frac{\alpha(r-1)[(2-\alpha)r-\alpha]}{4(r-\alpha)^2}\}(2\pi n\alpha(1-\alpha)^{-\frac{1}{2}}(\frac{(r-\alpha)^{r-\alpha}}{r^{r-2\alpha}\ \alpha^\alpha(1-\alpha)^{2(1-\alpha)}})^n$$

If one is an optimist, one may expect the variance of each of these r.v.s to be small. Thus $E(X_H^2) \sim E(X_H)^2$ would imply the conjecture about Hamilton cycles, and a stronger form of Corollary 3 would follow from $E(X_F^2) \sim E(X_F)^2$. Unfortunately the reality is not as beautiful as that. For example,

$$E(X_F^2) \sim e^{(r-1)/2}\ (\frac{r-1}{r-2})^{\frac{1}{2}}\ E(X_F)^2,$$

so for no $r \geq 3$ is the variance $o(E(X_F)^2)$.

In conclusion we give an application of the theory of random regular graphs to extremal graph theory. The maximal number of vertices in an independent set of a graph G is the *independence number* $\beta_o(G)$, and $\beta_o(G)/|G|$ is the *independence ratio*. An immediate consequence of a classical theorem of Brooks [26] is that if a graph has maximal degree $\Delta = \Delta(G)$ and does not contain a complete graph of order $\Delta+1$ then its independence ratio is at least $1/\Delta$. Lately several lower bounds have been given for the independence ratio under other "sparseness" conditions, e.g. in [2], [3], [46], [47], [50] and [75]. Perhaps the most natural measure of sparseness is the girth. Let $i(\Delta,g)$ be the infimum of the independence ratio of graphs with maximum degree Δ and girth at least g. Staton [75] proved that $i(\Delta,4) \geq 5/(5\Delta-1)$ and so an example of Fajtlowicz [46] implies that $i(3,4) = i(3,5) = 5/14$. Concerning cubic graphs of large girth, Hopkins and Staton [50] showed that $\lim_{g\to\infty} i(3,g) \geq 7/18$.

The results above say rather little about upper bounds for $i(\Delta,g)$. In particular, they do not exclude the rather natural possibility of $\lim_{g\to\infty} i(3,g) = \frac{1}{2}$. It turns out that these questions can be attacked by random regular graphs. In [17] the following result is proved.

Theorem 5.6. Let $\Delta \geq 3$ be a natural number and suppose that $0 < \alpha < 1$ satisfies

$$\alpha\{\Delta \log 2 + \log(1/\alpha)\} + (2-\alpha)(\Delta-1)\log(2-\alpha) + (\alpha+1)\Delta \log(1-\alpha)$$
$$< 2(\Delta-1)\log 2.$$

Then for every g there is a Δ-regular graph of girth at least g whose independence ratio is less than $\alpha/2$. In particular,

$$i(\Delta,g) < \alpha/2$$

and

$$i(3,g) < 6/13.$$

The theorem is proved by showing that almost every $G_{n,\Delta\text{-reg}}$ has independence ratio less than $\alpha/2$ while the probability of the girth being large is bounded away from zero.

REFERENCES

[1] M. Ajtai, J. Komlós and E. Szemerédi, A dense infinite Sidon sequence, European J. Combinatorics, to appear.

[2] M.O. Albertson, B. Bollobás and S. Tucker, The independence ratio and maximum degree of a graph, Prof. Seventh S-E Conf. Combinatorics, Graph Theory and Computing, Utilitas Math., Winnipeg, 1976, 43-50.

[3] M.O. Albertson and J.P. Hitchinson, The maximum size of an independent set in a nonplanar graph, Bull. Amer. Math. Soc. 81 (1975) 554-555.

[4] D. Angluin and L.G. Valiant, Fast probabilistic algorithms for Hamiltonian circuits and matchings, J. Computer and System Sciences 18 (1979) 155-193.

[5] L. Babai, P. Erdős and S.M. Selkow, Random graph isomorphisms, SIAM J. Comput. 9 (1980) 628-635.

[6] L. Babai and L. Kučera, Canonical labelling of graphs in linear average time, 20th Annual IEEE Symp. on Foundations of Comp. Sci. (Puerto Rico), 1979, 39-46.

[7] A. Békéssy, P. Békéssy and J. Komlós, Asymptotic enumeration of regular matrices, Studia Sci. Math. Hungar. 7 (1972) 343-353.

[8] E.A. Bender and E.R. Canfield, The asymptotic number of labelled graphs with given degree sequences, J. Combinatorial Theory (A) 24 (1978) 296-307.

[9] B. Bollobás, Extremal Graph Theory, Academic Press, London, New York and San Francisco, 1978.

[10] - , Chromatic number, girth and maximum degree, Discrete Math. 24 (1978) 311-314.

[11] - , Graph Theory - An Introductory Course, Graduate Texts in Mathematics, Springer-Verlag, New York, Heidelberg and Berlin, 1979.

[12] - , Degree sequences of random graphs, Discrete Math. 33 (1981) 1-19.

[13] - , A probabilistic proof of an asymptotic formula for the number of labelled regular graphs, Preprint Series 1979, Matematisk Institut, Aarhus Universitet.

[14] - , The distribution of the maximum degree of a random graph, Discrete Math. 32 (1980) 201-203.

[15] - , A probabilistic proof of an asymptotic formula for the number of labelled regular graphs, Europ. J. Combinatorics 1 (1980) 311-316.

[16] - , The diameter of random graphs, Trans. Amer. Math. Soc., to appear.

[17] - , The independence ratio of regular graphs, Proc. Amer. Math. Soc., to appear.

[18] - , Threshold functions for small subgraphs, to appear.

[19] - , Vertices of given degree in a random graph, J. Graph Theory, to appear.

[20] B. Bollobás and P. Catlin, Topological complete subgraphs of random graphs, J. Combinatorial Theory (B), to appear.

[21] B. Bollobás, P.A. Catlin and P. Erdős, Hadwiger's conjecture is true for almost every graph, European J. Combinatorics 1 (1980) 195-199.

[22] B. Bollobás and P. Erdős, Cliques in random graphs, Math. Proc. Cambridge Phil. Soc. 80 (1976) 419-427.

[23] B. Bollobás and N. Sauer, Uniquely colourable graphs with large girth, Canad. J. Math. 28 (1976) 1340-1344

[24] B. Bollobás and A.G. Thomason, Uniquely partitionable graphs, J. London Math. Soc. (2) 16 (1977), 403-410.

[25] C.E. Bonferroni, Teorie statistica delle classi e calcolo delle probabilità, Public. Inst. Sup. Sc. Ec. Comm. Firenze 8 (1936) 1-62.

[26] R.L. Brooks, On colouring the nodes of a network, Proc. Cambridge Phil. Soc. 37 (1941) 194-197.

[27] S.A.Burr, P. Erdős, R.J. Faudree, C.C. Rousseau and R.H. Schelp, An extremal problem in generalized Ramsey theory, Ars Combinatoria 10 (1980) 193-203.

[28] Yu. D. Burtin, On the probability of connectedness of a random subgraph of the n-cube, in Russian, Problemy Pered. Inf. 13 (1977).

[29] P.J. Cameron, On graphs with given automorphism group, Europ. J. Combinatorics 1 (1980) 91-96.

[30] P.A. Catlin, Hajós's graph-coloring conjecture: variations and counterexamples, J. Combinatorial Theory (B) 26 (1979) 268-274.

[31] K.L. Chung, A Course in Probability Theory, 2nd ed., Academic
Press, New York and London, 1974.

[32] G. Cornuejols, Degree sequences of random graphs, to appear.

[33] P. Erdös, Graph theory and probability, Canad. J. Math. 11
(1959) 34-38.

[34]   -   , Graph theory and Probability II, Canad. J. Math 13
(1961) 346-352.

[35] P. Erdös and S. Fajtlowicz, On the conjecture of Hajós, Tenth
S.E. Conf. on Combinatorics, Graph Theory and Computing, Boca Raton,
1979.

[36] P. Erdös and L.Lovász, Problems and results on 3-chromatic
hypergraphs and some related results, Infinite and Finite Sets
(A. Hajnal, R. Rado and V.T. Sós, eds.), Coll. Math. Soc. J.
Bolyai, vol. 11, Budapest, 1973, 609-627.

[37] P. Erdös and A. Rényi, On random graphs I, Publ. Math. Debrecen
6 (1959) 290-297.

[38]   -   , On the evolution of random graphs, Publ.
Math. Inst. Hungar. Acad. Sci. 5 (1960) 17-61.

[39]   -   , On the strength of connectedness of a
random graph, Acta Math. Acad. Sci. Hungar. 12 (1961),261-267

[40]   -   , On random matrices, Publ. Math. Inst.
Hungar. Acad. Sci. 8 (1964) 455-461.

[41]   -   , On the existence of a factor of degree
one of a connected random graph, Acta Math. Acad. Sci. Hungar. 17
(1966) 359-368.

[42]   -   , On random matrices II, Studia Sci. Math.
Hungar. 3 (1968) 459-464.

[43] P. Erdös and J. Spencer, Probabilistic methods in combinatorics
Academic Press, New York and London, 1974.

[44]   -   , Evolution of the n-cube, Comp. and
Maths. with Appls. 5 (1979) 33-39.

[45] P. Erdös and R.J. Wilson, On the chromatic index of almost all
graphs, J. Combinatorial Theory (B) 23 (1977) 255-257.

[46] S. Fajtlowicz, The independence ratio for cubic graphs, Proc.
Eighth S-E Conf. Combinatorics, Graph Theory and Computing,
Utilitas Math., Winnipeg, 1977, 273-277.

[47]   -   , On the size of independent sets in graphs, Proc.
Ninth S-E Conf. Combinatorics, Graph Theory and Computing, Utilitas
Math., Winnipeg, 1978, 269-274.

[48] R. Frucht, Graphs of degree 3 with given abstract group, Canad.
J. Math. 1 (1949) 365-378.

[49] G.R. Grimmett and C.J.H. McDiarmid, On colouring random graphs,
Math. Proc. Cambridge Phil. Soc. 77 (1975) 313-324.

[50] G.W. Hopkins and W. Staton, Girth and independence ratio, to
appear.

[51] G.I.Ivchenko, The strength of connectivity of a random graph,
Theory of Prob. and Appl. 18 (1973) 396-403.

[52] Ch. Jordan, Sur la probabilité des épreuves répetées, Bull. Soc.
Math. France 54 (1926) 101-137.

[53]   -   , Sur un cas généralisé de la probabilité des épreuves
répetées, Acta Scientiarum Math. (Szeged) 3 (1927) 193-210.

[54] Ch. Jordan, Le théorème de probabilité de Poincaré, généralisé au cas de plusieurs variables indépendantes, Acta Sci. Math. (Szeged) 7 (1934) 103-111.

[55] V. Klee and D. Larman, Diameters of random graphs, to appear.

[56] A.D. Korshunov, On the diameter of random graphs, Sovit Mat. Doklady 12 (1971) 302-305.

[57] - , Solution of a problem of Erdős and Rényi on Hamilton cycles in nonoriented graphs, Soviet Math. Doklady 17 (1976) 760-764.

[58] - , A solution of a problem of P. Erdős and A. Rényi about Hamilton cycles in non-oriented graphs (in Russian), Metody Diskr. Anal. v Teoriy Upr. Syst., Sbornik Trudov Novosibirsk 31 (1977) 17-56.

[59] D.W. Matula, On the complete subgraph of a random graph, Combinatory Mathematics and its Applications, Chapel Hill, N.C., 1970, 356-369.

[60] C. McDiarmid, Determining the chromatic number of a graph, SIAM J. Comput. 8 (1979) 1-14.

[61] - , Colouring random graphs badly, Graph Theory and Combinatorics (R.J. Wilson, ed.), Pitman Research Notes in Math. 34 (1979) 76-86.

[62] - , Clutter percolation and random graphs, Combinatorial Optimization (V.J. Rayward-Smith, ed.), Math. Programming Study, vol. 13, North-Holland, Amsterdam-New York-Oxford, 1980, 17-25.

[63] - , General percolation and random graphs, Advances in Applied Probability, to appear.

[64] J.W. Moon, Almost all graphs have a spanning cycle, Canad. Math. Bull. 15 (1972) 39-41.

[65] J.W. Moon and L. Moser, Almost all (0,1) matrices are primitive, Atudia Sci. Math. Hungar. 1 (1966) 153-156.

[66] G. Pólya, Kombinatorische Anzahlbestimmungen für Gruppen und Chemische Verbindungen, Acta Math. 68 (1937) 145-254.

[67] L. Pósa, Hamiltonian circuits in random graphs, Discrete Math. 14 (1976) 359-364.

[68] R.C. Read, The enumeration of locally restricted graphs (I), J. London Math. Soc. 34 (1959) 417-436.

[69] A. Schrijver, On the number of edge-colourings of regular bipartite graphs, A & E Report 18/80, University of Amsterdam.

[70] A. Schrijver and W.G. Valiant, On lower bounds for permanents, ZW 131/79, Mathematisch Centrum, Amsterdam.

[71] K. Schürger, Limit theorems for complete subgraphs of random graphs, Per. Math. Hungar. 10 (1979) 47-53.

[72] E. Shamir and E. Upfal, On factors in random graphs, Israel J. Math. (to appear).

[73] J. Spencer, Ramsey's theorem - a new lower bound, J. Combinatorial Theory (A) 18 (1975) 108-115.

[74] - , Asymptotic lower bounds for Ramsey functions, Discrete Math. 20 (1977) 69-76.

[75] W. Staton, Some Ramsey-type numbers and the independence ratio, Transactions Amer. Math. Soc. 256 (1979) 353-370.

[76] W.F. de la Véga, Long paths in random graphs, to appear.

[77]   E.M.Wright, Graphs on unlabelled nodes with a given number of edges, Acta Math. 126 (1971) 1-9.

[78]   -   , For how many edges is a graph almost certainly Hamiltonian? J. London Math. Soc. (2) 8 (1974) 44-48.

RECENT RESULTS IN GRAPH DECOMPOSITIONS

by

F. R. K. Chung and R. L. Graham
Bell Laboratories
Murray Hill, New Jersey

INTRODUCTION

The subject of graph decompositions is a vast and sprawling topic, one which we certainly cannot begin to cover in a paper of this length. Indeed, recently a number of survey articles and several books have appeared, each devoted to a particular subtopic within this domain (e.g., see [Fi-Wi], [Gr-Rot-Sp], [So 1], [Do-Ro]).

What we will attempt to do in this report is twofold. First, we will try to give a brief overall view of the landscape, mentioning various points of interest (to us) along the way. When possible, we will provide the reader with references in which much more detailed discussions can be found. Second, we will focus more closely on a few specific topics and results, usually for which significant progress has been made within the past few years. We will also list throughout various problems, questions and conjectures which we feel are interesting and/or contribute to a clearer understanding of some of the current obstacles remaining in the subject.

Notation

By a <u>graph</u>(1) G we will mean a (finite) set $V = V(G)$, called the *vertices* of G together with a set $E = E(G)$ of (unordered) pairs of vertices of G, called the *edges* of G.

Let H denote a family of graphs. By an *H-decomposition* of G we mean a partition of $E(G)$ into disjoint sets $E(H_i)$ such that each of the graphs H_i induced by the edge set $E(H_i)$ is isomorphic to a graph in H. Ordinarily we will just say that G has been decomposed into $H_i \in H$. If an H-decomposition of G exists, we denote this by writing $G \in <H>$.

By far the most work in graph decompositions has been carried out on the general problem of determining for fixed families H (usually singletons), necessary and sufficient conditions that $G \in <H>$. We begin by discussing several examples of this type.

Complete Graphs

For a fixed k, let H consist of the single graph K_k, the complete graph on k vertices. If $K_v^{(\lambda)}$ denotes the complete *multigraph* of multiplicity $\lambda \geq 1$ on a set of v vertices, i.e., each pair of vertices occurs as an edge exactly λ times, then the determination of necessary and sufficient conditions for

$$K_v^{(\lambda)} \in <\{K_k\}> \tag{1}$$

to hold is among the oldest problems in combinatories. Such a decomposition is easily seen to be equivalent to the existence of a (v,k,λ)-configuration, a combinatorial structure consisting of a family of k-element subsets B of a v-element set V in which every 2-element subset of V occurs in exactly λ of the B's (see [Hal] or [Ry]). It is not difficult to show that necessary conditions for (1) to hold are:

$$\lambda(v-1) \equiv 0 \pmod{k-1}; \tag{i}$$
$$\lambda v(v-1) \equiv 0 \pmod{k(k-1)}. \tag{ii}$$

For k = 3 and λ = 1 such configurations are known as Steiner triple systems. It was shown by Kirkman in 1847 that in this case (i) and (ii) are also sufficient for (1) (see e.g., [Ra-Wi 1] or the extensive bibliography in [Do-Ro]).

On the other hand, for the values $v = n^2 + n + 1$, $k = n + 1$, $\lambda = 1$, (1) holds iff there exists a projective plane of order n (PP(n)) (see [Ry]). In this case conditions (i) and (ii) are not sufficient since, for example, if such a plane exists then n must satisfy the celebrated condition of Bruck and Ryser [Bru-Ry], namely, $n \equiv 0$ or $1 \pmod 4$ and $n = x^2 + y^2$ for integers x and y. The first n for which the existence of PP(n) is undecided is n = 10. It has recently been shown that if PP(10) exists it must have very little symmetry (see [An-Hal]).

The strongest general result known for (1) is the theorem of Wilson [Wi 3].

Theorem. For fixed k and λ, (1) holds if (i) and (ii) hold, provided v is sufficiently large.

Thus, conditions (i) and (ii) are asymptotically sufficient.

More generally, let $I \subseteq \mathbb{Z}^+$, the positive integers, and consider the decomposition

$$K_v^{(\lambda)} \in <\{K_i : i \in I\}>. \tag{2}$$

In this case, Wilson [Wi 4] has proved an asymptotic result analogous to the preceding theorem. Let

$\alpha(I)$ denote g.c.d. $\{i-1 : i \in I\}$,

$\beta(I)$ denote g.c.d. $\{i(i-1) : i \in I\}$.

Theorem (Wilson). For fixed λ and $I \subseteq \mathbb{Z}^+$, if $v \geq v_0(I,\lambda)$ and v satisfies

$$\lambda(v-1) \equiv 0 \pmod{\alpha(I)} \tag{i'}$$
$$\lambda v(v-1) \equiv 0 \pmod{\beta(I)} \tag{ii'}$$

then (2) holds.

As before it is not hard to check that (i') and (ii') are *necessary* conditions for (2).

Typical results in this class of a more precise nature are:
Theorem (Hanani *[Han 2]*).

$K_v \in <\{K_3, K_4, K_6\}>$ iff $v \equiv 0$ or $1 \pmod 3$, $v \geq 3$,

$K_v \in <\{K_4, K_5, K_8, K_9, K_{12}\}>$ iff $v \equiv 0$ or $1 \pmod 4$, $v \geq 4$.

Theorem (Wilson *[Wi 2]*).

$K_v \in <\{K_3, K_5\}>$ iff $v \equiv 1 \pmod 2$, $v \geq 3$.

$K_v \in <\{K_4, K_7, K_{10}, K_{19}\}>$ iff $v \equiv 1 \pmod 3$, $v \geq 4$.

Theorem (Brouwer *[Br 3]*).

$K_v \in <\{K_4, K_7\}>$ iff $v \equiv 1 \pmod 3$, $v \geq 4$, $v \neq 10, 19$

$K_v \in <\{K_3, K_4, K_5, K_6, K_8\}>$ iff $v \geq 3$.

For a summary of many results of this type the reader is referred to the recent doctoral dissertation of Sotteau *[So 1]*.

If H consists of the set of *all* complete graphs then of course $G \in <H>$ for all G. In this case, however, it is of interest to know how many factors are required in the decomposition. In this case, Erdös, Goodman and Posa *[Er-Go-Po]* have shown that any graph on n vertices can be decomposed into at most $\lfloor n^2/4 \rfloor$ edge-disjoint complete graphs. In fact, they showed the same bound applies if one takes $H = \{K_2, K_3\}$ (and that this bound can be achieved).

If instead of minimizing the number of factors in a decomposition

$$E(G) = \sum_i E(K_{n_i}) \qquad (*)$$

we instead ask for the minimum value of

$$\sum_i v(K_{n_i}) = \sum_i n_i$$

then Chung *[Ch 4]* has shown that for any graph on n vertices, there is a decomposition (*) with

$$\sum_i n_i \leq \lceil n^2/2 \rceil,$$

settling an earlier conjecture of Katona and Tarján. Furthermore, the only graph for which the bound is achieved is $K_{\lfloor \frac{n}{2} \rfloor, \lceil \frac{n}{2} \rceil}$.

CYCLES AND PATHS

An extensive literature exists on decompositions of (complete) graphs into a fixed cycle C_k. Necessary conditions for

$$G \in <\{C_k\}> \qquad (3)$$

are:

$n - 1$ is even, $k \leq n$;   (iii)

$n(n-1) \equiv 0 \pmod{2k}$ (iv)

It is an old conjecture that these conditions are *sufficient* for (3) to hold but this is not yet known. Cases for which (iii) and (iv) *are* sufficient include the following:

(a) n = k, *[Ber]*

(b) k ≡ 0 (mod 4), *[Ko 1]*

(c) k ≡ 2 (mod 4), *[Ro 1]*

(d) n ≡ 0 (mod k), *[J 3]*

(e) n - 1 ≡ 0 (mod 2k), *[J 4]*

(f) k = 2p$^\alpha$, p prime, *[Al-Va]*.

The reader should consult *[So 1]* and *[Ga]* for a more complete discussion of known results.

In the case of triangles C_3 (=K_3), Nash-Williams has raised the following conjecture:

Conjecture [1]. If all vertices of G have even degrees at least $\frac{3}{4}$ v(G) then G ε <{C_3}>.

A variation of cycle (and complete graph) decompositions which has received some attention is that of a *resolvable* decomposition. In this case it is required that it be possible to partition the edge-disjoint cycles (partitioning E(G)) into classes, with the cycles in each class forming a partition of V(G). For example, the celebrated solution by Ray-Chaudhuri and Wilson *[Ra-Wi 1]* of the Kirkman school-girl problem shows that K_{6r+3} always has such a decomposition into C_3's. This general problem often is referred to as the "Oberwolfach" problem *[Gu]*. Partial results can be found in *[He-Ko-Ro]*, *[Hu-Ko-Ro]* and *[He-Ro]*.

Similar but less complete results are available for the case $K_n^{(\lambda)}$ ε <{C_k}>, λ > 1 (the reader should consult *[So 1]* for a summary). Of course, k = 3 is the previously mentioned case of Steiner triple systems.

Partial results for decompositions of complete multipartite graphs $K_{n,n,\ldots,n}$ into cycles are available in *[So 1]*, *[Co-Har]*, *[So 3]*. It is known for example, that

$$K_{r,s} \ \varepsilon \ <\{C_{2t}\}>$$

iff r ≡ s ≡ 0 (mod 2), r ≥ t, s ≥ t and rs ≡ 0 (mod 2t).

Relatively little is known for the case that *H* = <{C_i:iεI}> for a subset I ⊆ Z$^+$. The strongest conjecture as to what may be true is the following:

Conjecture (Alspach *[Al]*): Suppose n is odd and m_i ≥ 3, 1 ≤ i ≤ r, are integers satisfying

$$2 \sum_{i=1}^{r} m_i = n - 1.$$

Then

$$E(K_n) = \sum_{i=1}^{r} E(C_{m_i}).$$

If *H* = {C_i:i≥3} consists of *all* cycles then it is easy to see that G ε <*H*> iff all deg(v), v ε V(G), are even. In this case, however, there are a number of interesting conjectures concerning *minimal* cycle decompositions.

Conjecture (Hajos (see [Lo])). Every graph G on n vertices with all degrees even can be decomposed into at most $\left\lfloor \frac{n}{2} \right\rfloor$ edge-disjoint cycles.

If true the bound of $\left\lfloor \frac{n}{2} \right\rfloor$ would be best possible because of n.
The best result known in this direction is the theorem of Lovász [Lo]:

Theorem. If v(G) = n then G can be decomposed into at most $\left\lfloor \frac{n}{2} \right\rfloor$ edge-disjoint paths and cycles.

This is also the strongest partial result known towards the following beautiful question of Gallai:

Conjecture: If v(G) = n then G can be decomposed in at most $\left\lfloor \frac{n}{2} \right\rfloor$ paths.

Suppose $\ell(G)$ denotes the minimum number of *linear forests* (= union of paths) into which a graph G can be decomposed.

Conjecture ([Pe] and [Ak-Ex-Har]).

$$\ell(G) \leq \left\lceil \frac{1}{2} (\Delta(G)+1) \right\rceil$$

where $\Delta(G)$ denotes the maximum degree in G.

In [Pe] it is shown that

$$\left\lceil \frac{1}{2}\Delta(G) \right\rceil \leq \ell(G) \leq \lceil (2\Delta(G)+1)/3 \rceil$$

so that if valid, the bound in the conjecture would be close to best possible.

We mention in passing that the special case of $\ell(G)$ in which each of the paths is required to be a single edge has been intensively studied. In this case $\ell(G)$ is usually denoted by $\chi'(G)$ and called the *chromatic index* of G. It is just the minimum number of colors needed for coloring the edges of G so that neighboring edges have distinct colors. The reader is referred to the excellent monograph of Fiorini and Wilson [Fi-Wi] which summarizes much of what is currently known about $\chi'(G)$. The well known theorem of Vizing asserts that $\chi'(G) = \Delta(G)$ or $\Delta(G) + 1$ for all graphs G (where $\Delta(G)$ denotes the maximum degree of G). It is known [Er-Wi] that almost all graphs G have $\chi'(G) = \Delta(G)$ although Holyer [Hol] has shown that the problem of determining whether $\chi'(G) = \Delta(G)$ or $\Delta(G) + 1$ is NP-complete.

Along somewhat different lines, the following old conjecture of S. Lin to the best of the authors' knowledge still remains unsettled.

Conjecture (Lin [Lin]). If v(G) = n, e(G) > n and $G \in \langle\{C_n\}\rangle$ then G has at least two distinct representations as $\sum E(C_n)$.

No doubt the number of such representations is bounded below by some increasing function of e(G).

It has been noted by Sloane [Sl] (using a result of Tutte [Tu]) that the edge disjoint union of any two hamiltonian cycles of a graph always contains a third hamiltonian cycle.

A related conjecture of Nash-Williams is still open:

Conjecture [5]. If deg(v) = 2k for all v ε V(G) and v(G) ≤ 4k then G can be decomposed into at most k hamiltonian cycles.

It has been recently shown by Jackson *[J 1]* that under these hypotheses G always contains at least $\left\lceil \frac{k}{3} \right\rceil$ edge-disjoint hamiltonian cycles.

Kotzig *[Ko 2]* has proposed the following related

Conjecture: If G_i can be decomposed into p_i hamiltonian cycles, $1 \le i \le r$, then the cartesian product $\prod_{i=1}^{r} G_i$ can be decomposed into $\sum_{i=1}^{r} p_i$ hamiltonian cycles.

This was shown to be true for $p_1 = p_2 = 1$ by Kotzig *[Ko 2]*, $p_1 = p_2 = p_3 = 1$ by Foregger *[For]* and r = 2, $p_2 \le p_1 \le 2p_2$, by Aubert and Schneider *[Au-Schn]*.

For further discussions of these and related questions (especially the analogues for directed graphs) the reader is referred to *[J 2]*.

A SINGLE ARBITRARY GRAPH

Suppose *H* = {H} where we assume (as usual) that H contains no isolated vertex. Let us denote by deg(x) the *degree* of a vertex in a graph. Define D(H) by

$$D(H) \equiv \sum_{v \in V(H)} z_v \deg(v) : z_v = 0,1,2,\dots \ ,$$

i.e., D(H) is the set of nonnegative integer linear combinations of the deg(v), v ε V(H).

As before the obvious necessary conditions for

G ε <{H}>   (4)

can be easily stated:

e(G) ≡ 0 (mod e(H));   (v)

For each vertex x ε V(G), deg(x) ε D(H).   (vi)

As in previous cases, most of the work on this problem has been concerned with the choices $G = K_v^{(\lambda)}$, and in particular, $G = K_v$. The strongest asymptotic results are given by the following beautiful theorem of Wilson:

Theorem (Wilson *[Wi 2]*, *[Wi 4]*).

For all λ > 0 and H there exists (a least) v(H,λ) so that if:

(a) v ≥ v(H,λ),

(b) λv(v-1) ≡ 0 (mod 2e(H)),

(c) λ(v-1) ≡ 0 (mod d) where d = g.c.d. {deg(v):vεV(H)},

then

$K_v^{(\lambda)}$ ε <{H}>

In other words, for $K_v^{(\lambda)}$ the necessary conditions (v), (vi) are sufficient for v sufficiently large.

A detailed analysis of such decompositions of K_v for each graph H with v(H) ≤ 5 has been carried out by Bermond, Huang, Rosa and Sotteau

[Berm-Hu-Ro-So], extending earlier work of Bermond and Schonheim
[Berm-Sc] who treated all H with $v(H) \leq 4$. In general they found
that the exact values of $v(H,1)$ they obtained were always rather small,
in particular, much smaller than the bounds implicit in the constructions of Wilson.

A variation which has received some attention recently is to
decide whether a graph G occurs nontrivially in <{H}> for *any* graph
H (where nontrivial means $G \neq H$). From an algorithmic point of view,
it has been shown by Graham and Robinson [Gr-Rob] that the problem is
NP-complete, even if G is a tree and a decomposition into two isomorphic
subgraphs is required (see [Gar-Jo] for a discussion of NP-completeness).

For $G = K_v$, it was shown by Harary, Robinson and Wormald [Hara-Rob-
Wo 1] that the necessary condition for the decomposition of K_v into t
isomorphic edge-disjoint subgraphs is sufficient, namely,

$v(v-1)/2 \equiv 0 \pmod{t}$.

Similar results for other classes of graphs (including directed graphs)
can be found in [Hara-Rob-Wo 2], [Hara-Rob-Wo 3], [Wa].

TREES

The following conjecture is usually attributed to Ringel [Ri] (see
[Ro 2]).

Conjecture: For any tree T with $e(T) = n$,

$K_{2n+1} \in <\{T\}>$.

Kotzig strengthened Ringel's conjecture and conjectured that every
K_{2n+1} has a *cyclic* decomposition into trees isomorphic to a fixed tree
T with $e(T) = n$. This is equivalent to asserting that every tree T is
graceful, i.e., there exists a 1 - 1 labelling $\lambda : V(T) \to \{0,1,\ldots,e(T)\}$
such that all the values $|\lambda(i)-\lambda(j)|$, $e = \{i,j\} \in E(T)$ are distinct.
Although still unresolved, this conjecture has stimulated numerous
papers dealing with various special cases. A discussion of much of this
work can be found in the survey papers of Bloom [Bl], and Huang, Kotzig
and Rosa [Hu-Ko-Ro 2].

An analogous concept, that of a *harmonious* graph, in which each
edge $\{i,j\}$ is assigned the value $\lambda(i) + \lambda(j)$ modulo $e(G)$ and all edge
values are required to be distinct, has been studied recently. The
connection of this concept with coding theory and additive number
theory is covered in [Gr-Sl 1] and [Gr-Sl 2]. A particularly stubborn
problem is the following.

Problem: Is it true that for an absolute $\varepsilon > 0$, every harmonious graph
G with n vertices must have

$$e(G) < \left[\frac{1}{2} - \varepsilon\right]n^2?$$

In other words, if $\{a_1,\ldots,a_n\} \subseteq \mathbb{Z}_e$ has the property that every element $z \in \mathbf{Z}_e$ can be written as $z \equiv a_i + a_j$ (mod e) then is it true that $e < \left[\frac{1}{2} - \epsilon\right] n^2$?

It is known that harmonious graphs with n vertices and $\frac{5}{18}(1+o(1))n^2$ edges exist. Also, it is not hard to show that almost all graphs are neither graceful nor harmonious.

In [Ch 3], Chung considers the problem of decomposing a connected graph G into a minimum number $\tau(G)$ of trees. She shows that at most $\left\lceil\frac{v(G)}{2}\right\rceil$ are ever required and that this bound is achieved, for example, for complete graphs. A related result of Nash-Williams [Nash 1] proves that the minimum number of *forests* (i.e., acyclic graphs) a graph G can be decomposed into is exactly

$$\max\left\{\frac{e(H)}{v(H)-1}: \text{ H is an induced subgraph of } G\right\}.$$

(Related work occurs in [Re],[Bei],[Ak-Ha]).

The consideration of $\tau(G)$ was suggested by results of M. and T. Forreger who considered the related quantity $\tau'(G)$, defined to be the minimum number of subsets into which V(G) can be partitioned so that each subset induces a tree. They show [For-For] that

$$\tau'(G) \leq \left\lceil\frac{1}{2}v(G)\right\rceil.$$

The relationship between $\tau(G)$ and $\tau'(G)$ is not yet completely understood. Examples are known for which

$$\frac{\tau(G)}{\tau'(G)} > \frac{1}{8}v(G)$$

and

$$\frac{\tau'(G')}{\tau(G')} > \frac{1}{4}v(G').$$

These are probably not the extreme values these ratios can achieve.

Question: What are the extreme values of $\frac{\tau(G)}{\tau'(G)v(G)}$ and $\frac{\tau'(G)}{\tau(G)v(G)}$?

An extended discussion for decompositions of K_n into trees is given by Huang and Rosa in [Hu-Ro 2]. In particular they determine which $K_n \in \langle\{T\}\rangle$ for all trees with $e(T) \leq 8$.

In this connection, Gyarfas and Lehel have raised the following striking conjecture:

Conjecture [Gy-Le]. If $\{T_i:1\leq i\leq n-1\}$ is an arbitrary set of trees with $e(T_i) = i$ then $E(K_n)$ can be decomposed into $\sum_{i=1}^{n-1} E(T_i)$.

This is known to hold, for example, if all the T_i are either stars or paths [Za-Lui], [St]. However, at present it is not even known that the *degree sequences* of the T_i can always be arranged so that the (vector) sum is the degree sequence of K_n, i.e., $(n-1,\ldots,n-1)$.

COMPLETE BIPARTITE GRAPHS

Decompositions of graphs G into complete bipartite subgraphs behave in a somewhat different manner than for other graphs. One reason for this, for example, is the fact that the *spectrum* of G (i.e., the set of eigenvalues of the adjacency matrix of G), strongly limits the minimum number of complete bipartite factors in such a decomposition of G. More precisely, if $n^+(G)$ and $n^-(G)$ denote the numbers of positive and negative eigenvalues, respectively, of A(G) (which always has all eigenvalues real), then

$$E(G) = \sum_{i=1}^{t} E(K_{r_i,s_i})$$

implies

$$t \leq \max\{n^+(G), n^-(G)\}$$

(see [Gr-Pol] or [Ho]). No analogous bounds can exist for decompositions into complete graphs.

Very little work has been done for the general decomposition problem $G \in <\{K_{r,s}\}>$. The cases in which $(r,s) = (2,2)$ and $(2,3)$ are treated in [Hu] and $(r,s) = (2,4)$ and $(3,3)$ are treated in [Hu-Ro 1] under the additional restriction that the decomposition be *balanced*, i.e., each vertex of G appears in the same number of factors. This type of restriction on general compositions has been investigated in a number of papers (see [So 1] for a discussion of this work).

An interesting variation of decomposition into complete bipartite graphs has recently been considered by Chung, Erdős and Spencer [Ch-Er-Sp]. Define the function $\alpha(n)$ to be the least integer such that any graph G on n vertices can be decomposed into complete bipartite subgraphs

$$E(G) = \sum_i E(K_{r_i,s_i})$$

with

$$\sum_i v(K_{r_i,s_i}) = \sum_i r_i s_i \leq \alpha(n)$$

Theorem

$$\frac{\alpha(n) \log n}{n^2} = O(1).$$

In order to provide the reader with an idea for the type of techniques useful in estimates of this kind (and since the results are not available in the literature) we sketch a proof.

We first show

$$\alpha(n) \geq (1-\epsilon) \frac{n^2}{2e \log n} \tag{5}$$

for any $\epsilon > 0$ and all sufficiently large n. Consider a random[2] graph G with n vertices and $\lfloor n^2/2e \rfloor$ edges. The probability that G contains $K_{a,b}$ is bounded above by

$$\binom{n}{a}\binom{n}{b} e^{-ab} < e^{(a+b)\log n - ab}.$$

Let S denote the set of all unordered pairs $\{a,b\}$ satisfying

$$1 \le a, b \le n \text{ and } \frac{a+b}{ab} < \frac{1-\varepsilon}{\log n}$$

The probability that G contains a $K_{a,b}$ with $\frac{a+b}{ab} < \frac{1-\varepsilon}{\log n}$

is bounded above by

$$\sum_{\{a,b\}\varepsilon S}\binom{n}{a}\binom{n}{b} e^{-ab} < \sum_{\{a,b\}\varepsilon S} e^{-\varepsilon ab}$$

$$< \sum_{\{a,b\}\varepsilon S} e^{-\varepsilon \log^2 n}$$

$$< n^2 e^{-\varepsilon \log^2 n} < 1$$

for large n. Thus, there exists a graph G with n vertices an $\lfloor n^2/2e \rfloor$
edges which does not contain any such $K_{a,b}$ as a subgraph.

Let

$$E(G) = \sum_i E(K_{r_i,s_i})$$

be a decomposition of G having $\sum(r_i+s_i)$ minimal. For any edge $\{u,v\}$
in G, define

$$f(u,v) = \frac{r_i+s_i}{r_i s_i}$$

where $\{u,v\} \varepsilon K_{r_i,s_i}$. Then

$$\sum_i(r_i+s_i) = \sum_{\{u,v\}} f(u,v).$$

By hypothesis, any K_{r_i,s_i} occurring in the decomposition has

$$\frac{r_i+s_i}{r_i s_i} \ge \frac{1-\varepsilon}{\log n}.$$

Thus

$$f(u,v) \ge \frac{1-\varepsilon}{\log n}$$

for any $\{u,v\} \varepsilon E(G)$ and consequently

$$\alpha(n) \ge \frac{(1-\varepsilon)n^2}{2e \log n}$$

which proves (5).

We next show

$$\alpha(n) < (1+\varepsilon) \frac{n^2}{2 \log n} \tag{6}$$

for any $\varepsilon > 0$ and all n sufficiently large. We first need the following:

Proposition. For any $\varepsilon > 0$ and $\rho > 0$, any graph G on n vertices and
$\rho\binom{n}{2}$ edges contains a subgraph isomorphic to $K_{r,s}$ for some r, s with
$r > \varepsilon \rho n$ and $s > (1-\varepsilon)\rho^r n$.

Proof of Proposition: Suppose $v(G) = n$, $e(G) \ge \rho\binom{n}{2}$ and G does not
contain $K_{r,s}$ as a subgraph where $r = \lceil \varepsilon \rho n \rceil$ and $s = \lceil(1-\varepsilon)\rho^r n\rceil$. Thus, if

we consider the adjacency matrix $A(G) = (a_{ij})$ of G then

$$\sum_{j=1}^{n} \sum_{1 \le i_1 < \ldots < i_r \le n} a_{i_1,j} \ldots a_{i_r,j} \le (s-1) \binom{n}{r}. \tag{6}$$

The left-hand side of (6) is minimized by choosing all n of the sums $\sum_{i=1}^{n} a_{ij}$ as equal possible.

Thus, since

$$\sum_{j=1}^{n} \sum_{i=1}^{n} a_{ij} = 2e(G) = \rho n(n-1)$$

then

$$(s-1) \binom{n}{r} \ge \binom{\rho(n-1)}{r} n. \tag{7}$$

However, this is incompatible with the assigned values of r and s.

Continuing the proof of (6), the proposition guarantees that a graph G_0 on n vertices and $\rho\binom{n}{2}$ edges contains a subgraph H_0 isomorphic to K_{r_0,s_0} where

$$r_0 = \lfloor (1-\varepsilon_0) \log n/\log (1/\rho) \rfloor,$$

$$s_0 = \lfloor (r_0^2/\log (1/\rho) \rfloor.$$

where $\varepsilon_0 > \frac{2 \log \log n}{\log n}$. We will decompose G_0 into complete bipartite subgraphs by a "greedy" algorithm. Given G_0, we find a subgraph H_0 isomorphic to K_{r_0,s_0} and we let G_1 be the subgraph of G_0 with edge set $E(G_1) = E(G_0) - E(H_0)$. Next, we find a subgraph H_1 isomorphic to K_{r_1,s_1} and we let G_2 be the subgraph of G_1 with edge set $E(G_1) - E(H_1)$. We continue in this fashion until at most $\varepsilon_1 n^2/\log n$ edges remain. Therefore,

$$E(G_0) = \sum_{i \ge 0} E(K_{r_i,s_i}) \cup S$$

where S is a set of at most $\varepsilon_1 n^2/\log n$ edges.

We will prove by induction on the number of edges that for given ε_1, $0 < \varepsilon_1 < \varepsilon_0$, and n sufficiently large

$$\sum_{i \ge 0} (r_i + s_i) \le (1+\varepsilon_1) \frac{n^2}{2 \log n} \int_0^\rho \log (1/x) dx + 2\varepsilon_1 n^2/\log n \tag{8}$$

Since

$$\sum_{i \ge 0} (r_i + s_i) = r_0 + s_0 + \sum_{i \ge 1} (r_i + s_i)$$

then by induction

$$\sum_{i \geq 0} (r_i + s_i) + |S| \leq \frac{(1-\varepsilon_1)(\log n)^2}{(\log(1/\rho))^3} + (1+\varepsilon_1) \frac{n^2}{2 \log n} \int_0^{\rho'} \log(1/x)dx$$

$$+ 2\varepsilon_1 n^2/\log n$$

where $\rho' = (e(G_0) - r_0 s_0)/\binom{n}{2}$ and n is sufficiently large. However, straightforward calculation shows that

$$\frac{(1-\varepsilon_1)(\log n)^2}{(\log(1/\rho))^3} + (1+\varepsilon_1) \frac{n^2}{2 \log n} \int_0^{\rho'} \log(1/x)dx$$

$$\leq (1+\varepsilon_1) \frac{n^2}{2 \log n} \int_0^{\rho} \log(1/x)dx + 2\varepsilon_1 \frac{n^2}{\log n}$$

Thus,

$$\alpha(n) \leq (1+\varepsilon_1) \frac{n^2}{2 \log n} \int_0^1 \log(1/x)dx + 2\varepsilon_1 \frac{n^2}{\log n}$$

$$\leq (1+\varepsilon) \frac{n^2}{2 \log n}$$

for any preassigned $\varepsilon > 0$, provided n is sufficiently large. This proves (6).

The theorem follows by combining (5) and (6). □

We mention in passing the following:

Problem. Find an explicit construction for a graph G on n vertices and cn^2 edges (or even $cn^2/\log n$ edges) which contains no $K_{m,m}$ as a subgraph with $m = c' \log n$.

H-FREE GRAPHS

At the other end of the spectrum, a large number of papers have appeared within the last 10 years which deal with the following question. For a fixed graph H, we say that a graph G is *H-free* if G contains no subgraph isomorphic to H. Define $\alpha(G;\bar{H})$ to be the minimum number of factors possible in an H-free decomposition of G.

(BIG) PROBLEM

Determine (or estimate) $\alpha(G;\bar{H})$ for various families of G and H.

When G is a complete graph then $\alpha(G;\bar{H})$ is what might be thought of as an inverse "Ramsey number". In particular, if $r(H;k) = r$ denotes the least integer so that any k-coloring of the edges of K_r always forms a monochromatic subgraph isomorphic to H, then

$$\alpha(K_{r(H;k)};\bar{H}) = k + 1$$

There are a rather large number of recent survey papers covering this interesting topic, e.g., [Bu 1], [Bu 2], [Be-Ch-Le], [Par], [Gr],

[Bo], [Gr-Rot-Sp], [Ne-Rod]. Rather than duplicate their contents, we will restrict ourselves to mentioning several of what we consider to be the most attractive open problems in the area.

Question (Erdős). Does $\lim\limits_{n \to \infty} r(K_n;2)^{1/n}$ exist? If so, what is its value?

(It is known that it must be between $\sqrt{2}$ and 4 (see [Gr-Rot-Sp] or [Er-Sp]).

Question [Er-Gr]. Is it true that if T_m is a tree with m vertices then for fixed k, $r(T_m;k) = (1+o(1))mk$? It is known [Er-Gr] that it lies between $\frac{1}{2}(1+o(1))mk$ and $2(1+o(1))mk$.

Question: Is $\lim\limits_{k \to \infty} r(K_3;k)^{1/k} < \infty$?

It is known [Ch 2] that the limit exists and is greater than 3.1 (see [Ch 1]).

Define a family G of graphs to be *L-set* if for some absolute constant $c = c(G)$,

$r(G;2) \le cv(G)$ for all $G \in G$.

Define the (local) edge density $\rho(G)$ of a graph G by

$$\rho(G) = \max_{H \subseteq G} \frac{e(H)}{v(H)}.$$

(Strong) Conjecture (Erdős). If $\rho(G)$ is bounded for $G \in G$ then G is an L-set.

Conjecture (Erdős). If G_m has chromatic number m then $r(G_m;2) \ge r(K_m;2)$.

Question: Is it true that if H is any C_4-free graph then for any k there exists another C_4-free graph G_k so that $\alpha(G_k;H) > k$?

This is known [Ne-Rod] to be true for K_m-free graphs and C_{2m+1}-free graphs.

Problem [Er-Fa-Ro-Sc]. If $\alpha(G;\overline{P}_n) > 2$ then how small can e(G) be (where P_n denotes the path of length n)?

It is rather embarrassing that at present we can rule out neither $e(G) > cn^2$ nor $e(G) < cn$!

There are many other beautiful problems still open in Ramsey graph theory which unfortunately we must restrain ourselves (because of space limitations) from discussing. Many can be found in [Ne-Rod], [Gr], [Bu 1], [Bu 2].

We close this section with a final problem which has been annoying a number of people for (what seems to us to be) an unreasonable length of time. Let c denote the set of odd cycles. It is not hard to see that

$$\alpha(K_{2^n};\overline{c}) = n, \quad \alpha(K_{2^n+1};\overline{c}) > n.$$

In other words, it is possible to decompose K_{2^n} into n *bipartite* graphs but this is not possible for K_{2^n+1}. Let L(n) denote the least integer such that in every decomposition of K_{2^n+1} into n subgraphs, some sub-graph has an odd cycle of length at most L(n).

Question: Does $L(n) \to \infty$ as $n \to \infty$?

SIMULTANEOUS DECOMPOSITION

Given two graphs G and G' with e(G) = e(G'), by a *U-decomposition* of *G and G'* we mean a pair of partitions $E(G) = \sum_{i=1}^{r} E_i$, $E(G') = \sum_{i=1}^{r} E_i'$, such that as graphs, E_i and E_i' are isomorphic for all i. The function $U(G,G')$ is defined to be the minimum value of r for which a U-decomposition of G and G' into r parts exists. U-decompositions always exist when e(G) = e(G') since we can choose all the E_i and E_i' to be single edges.

A number of standard graph invariants can be placed into this frame-work. For example, if G' consists of e(G) disjoint edges then U(G,G') is just the chromatic index of G (mentioned earlier). When G' is a *star* of degree e(G) then U(G,G') is known as the edge-dominating number of G. Similarly, min U(G,G') has been called the thickness, arboricity or
                     G'
biparticity of G (see *[Har]*, *[Har-Hs-Mi]*) when G' ranges over all planar graphs, acyclic graphs or bipartite graphs, respectively.

Several recent papers have dealt with the quantity
$$U(n) \equiv \max_{G,G'} U(G,G')$$
where v(G) = v(G') = n (and, of course, e(G) = e(G')). The basic result is this.

Theorem [Ch-Er-Gr-Ul-Ya].
$$U(n) = \frac{2}{3} n + o(n).$$

The bound is achieved by (approximately) taking G to be a star of degree n and G' to be $\frac{n}{3}$ disjoint triangles.

A rather surprising phenomenon occurs for the analogous function $U_m(n)$, defined by simultaneously decomposing k graphs $G_1,\ldots,G_k$ into mutually isomorphic subgraphs (so that $U(n) = U_2(N)$). In this case:

Theorem [Ch-Er-Gr 1]. For all $k \geq 3$,
$$U_k(n) = \frac{3}{4} n + o(n)$$
where the o(n) term depends only on k.

It was completely unexpected that the coefficient $\frac{3}{4}$ would be independent of k, for $k \geq 3$.

Three graphs which drive $U(G_1,G_2,G_3)$ up to $\frac{3}{4}$ n are:

G_1 = a star of degree n;

$G_2 = \frac{n}{3}$ K_3's;

$G_3 = \left(\frac{n-\sqrt{n}}{2}\right)$ disjoint edges together with $K_{\sqrt{n}}$.

If the graphs under consideration are restricted to be bipartite then corresponding function $U_k^*(n)$ satisfies:

$$U_2^*(n) = \frac{n}{2} + o(n),$$

$$U_k^*(n) = \frac{3}{4} n + o(n).$$

L.Babai has raised the following tantalizing question.

Question: Is it true that if for some $\varepsilon > 0$, $v(G) = v(G') = n$ and $e(G) = e(G') > \varepsilon n^2$ then $U(G,G') = o(n)$?

(Weak) supporting evidence for an affirmative answer is that all known examples for achieving $U(G,G') = \frac{3}{4} n + o(n)$ have, in fact, a *linear* number of edges.

The question "Is $U(G,G') = 1?$," known as the graph isomorphism problem, has been actively studied recently from an algorithmic point of view. It is known (see [Lu], [Ba]) that for any fixed bound on $\Delta(G)$, there is a polynomial time algorithm (in $v(G)$) for testing isomorphism to G. The general problem for arbitrary graphs has not been shown to be NP-complete (and, many researchers feel that it is not). However, the related question "Is $U(G,G') = 2?$" has been proved by F. Yao [Ya] to be NP-complete.

For trees T, T', the question "Is $U(T,T') = k?$" has a polynomial time solution for $k = 1$ and is undecided for $k > 1$.

OTHER DIRECTIONS

In this final section we indicate some of the variations on our central theme which can be found in the literature.

Suppose H_2 denotes the class of all graphs of diameter exactly 2. In [Bos-Er-Ro], Bosák, Erdős and Rosa show (among other things) that for any $k > 2$, there is a complete graph $K_{r(k)} \in \langle H \rangle$ which has a decomposition with exactly k factors (extending earlier work of [Bos-Ro-Zn]). A number of papers (e.g., [Zn 1], [Zn 2], [Pa], [To]) have dealt with questions of a similar type for graphs of diameter d, especially in the case that all the factors are required to be isomorphic (see [Ko-Ro] for a survey of these results).

Many of the problems and/or results described in earlier sections have *directed* analogues. For example, suppose K_n^* denotes the complete symmetric directed graph on n vertices. A necessary condition that $K_n^* \in \langle \{C_k^*\} \rangle$ (where C_k^* denotes a directed k-cycle) is that $n \geq k$ and $n(n-1) \equiv 0 \pmod{k}$.

Conjecture (Bermond): These conditions are sufficient for $K_n^* \in \langle \{C_k^*\} \rangle$ except for $(n,k) = (4,4)$, $(6,3)$ and $(6,6)$.

Sotteau [So 2] has shown, for example, that if $k \geq 5$ is odd, $n \geq k$ and $n \equiv 0$ or $1 \pmod{k}$ then $K_n^* \in \langle \{C_k^*\} \rangle$, settling the above conjecture when k is an odd prime power, or k is odd and n is a prime power. Many references to this and related work can be found in [So 1].

In another direction, one might ask the analogous questions for hypergraphs (and indeed, people have). Typical results range from the difficult area of t-designs (see, e.g., [Ra-Wi 3], [Gr-Li-L], [Wi 1], [Br 1]), the beautiful theorem of Baranyai on complete hypergraph decompositions, [Bar], [Ca], (and more generally, hypergraph designs (see [Br-Sch] for a survey) directed hypergraph decompositions [Ge], and U-decompositions of hypergraphs [Ch-Er-Gr 2], to name a few. An especially stubborn problem of this type (and one for which Erdős is offering US $500) is the following problem.

Problem. (Erdős, Faber, Lovász [Er]). Suppose F is a family of n n-sets such that for any F, F' ε F, F $\neq$ F', we have $|F \cap F'| \leq 1$. Is it true that it is always possible to partition the underlying set $\underset{F \varepsilon F}{\cup}$ of vertices into n classes $C_1, \ldots C_n$ so that $|C_i \cap F| \leq 1$ for all i and all F ε F?

Some partial results can be found in [Hi].

In most of these variations, one might also ask when *resolvable* decompositions are possible, i.e., so that the vertex sets of the factors can be grouped to form partitions of the factored graph. This topic also has a wide literature, some of which can be found in [Han-Ray-Wi], [Han 1], [Ra-Wi 2], [J 2], [Ka], and especially [So 1].

FOOTNOTES

1. Usually we will adopt the graph-theoretic terminology in [Har].
2. For a detailed treatment of the use of the probabilistic method, see [Er-Sp].

REFERENCES

[Ak-Ha]   J. Akiyama and T. Hamada, A note on the arboricity of the complement of a tree, TRU Math, 13 (1977), 55-58.

[Ak-Ex-Ha]   J. Akiyama, G. Exoo and F. Harary, Covering and packing in graphs IV: linear arboricity, preprint.

[Al]   B. Alspach (personal communication).

[Al-Va]   B. Alspach and B. Varma, Cycles and graph decompositions, Proc. Joint Canada-France Colloq. on Combinatorics, Annals of Dis. Math. (to appear).

[An-Hal]   R. P. Anstee and M. Hall, Jr., Planes of order 10 do not have a collineation of order 5, J. Combinatorial Theory (A), 29 (1980), 39-58.

[Au-Schn]   J. Aubert and B. Schneider, Decomposition de la somme cartésienne d'un cycle et de l'union de deux cycles hamiltonien en cycles hamiltoniens, Research report No. 452 Lab de Recherche en Informatique, Univ. Paris-Sud 1980.

[Ba]   L. Babai, On the complexity of canonical labeling of strongly regular graphs, SIAM J. on Computing, Vol. 9 (1980), 212-216.

[Bar]   Z. S. Baranyai, On the factorization of the complete uniform hypergraph, Infinite and Finite Sets (Colloq., Keszthely, 1973, Vol. I), Colloq. Math. Soc. János Bolyai, 10, North Holland, Amsterdam (1975).

[Be-Ch-Le]   M. Behzad, G. Chartrand and L. Lesniak-Foster, Graphs and Digraphs, Prindle, Weber and Schmidt, Boston, Mass. 1979.

[Bei]   L. W. Beineke, Decompositions of complete graphs into forests, MTA Mat. Kut. Int. Közl 9 (1964), 589-594.

[Ber]   C. Berge, Graphs and Hypergraphs, North Holland, Amsterdam, 1973.

[Berm-Ge-So]   J. C. Bermond, A. Germa and D. Sotteau, Hypergraph-designs, Ars Combinatoria, 3 (1977), 47-66.

[Berm-Hu-Ro-So]   J. C. Bermond, C. Huang, A. Rosa and D. Sotteau, Decomposition of complete graphs into isomorphic subgraphs with five vertices, Ars Combinatoria, (to appear).

[Berm-Sc]   J. Bermond and J. Schonheim, G-decomposition of K_n, where G has four vertices or less, Dis. Math., 19 (1977),113-120.

[Bl]   G. S. Bloom, A chronology of the Ringel-Kotzig conjecture and the continuing quest to call all trees graceful, in Topics in Graph Theory, Annals of N. Y. Acad. Sci., 328 (1979), 32-51.

[Bo]   B. Bollobás, Graph Theory, Springer, Berlin, 1979.

[Bos-Er-Ro]   J. Bosák, P. Erdős and A. Rosa, Decompositions of complete graphs into factors with diameter two, Mat. Časopis, 21 (1971), 14-28.

[Bos-Ro-Zn]   J. Bosák, A. Rosa and Š. Znám, On decompositions of complete graphs into factors with given diameter, Theory of graphs into factors with given diameter, Theory of Graphs, Proc. Colloq. Tihany 1966, Publ. House Hungarian Acad. Sci., Budapest (1968), 37-56.

[Br 1]   A. E. Brouwer, The t-designs with v < 18, Math. Centre report ZN 76, August 1977.

[Br 2]   -   , Wilson's Theory, in Mathematical Centre Tracts, 106, 75-88, Mathematisch Centrum, Amsterdam, 1979.

[Br 3]   -   , (personal communication).

[Br-Sch]   A. E. Brouwer and A. Schrijver, Uniform hypergraphs, in Mathematical Centre Tracts, 106 (1979), 39-73, Mathematisch Centrum, Amsterdam, 1979.

[Bru-Ry]   R. H. Bruck and H. J. Ryser, The nonexistence of certain finite projective planes, Canad. Jour. Math, 1 (1949), 88-93.

[Bu 1]   S. A. Burr, Generalized Ramsey theory for graphs - A survey, In Graphs and Combinatorics, Springer, Berlin, 1974, 52-75.

[Bu 2]   -   , A survey of noncomplete Ramsey theory for graphs, in Topics in Graph Theory, Annals of N.Y.

[Ca]   P. J. Cameron, Parallelisms of Complete Designs, Cambridge Univ. Press, New York 1976.

[Ch 1]   F. R. K. Chung, On the Ramsey numbers N(3,3,...,3;2), Dis. Math., 5 (1973), 317-321.

[Ch 2]   -   , On triangular and cyclic Ramsey numbers with k colors, In Graphs and Combinatorics, Springer, Berlin, 1974, 236-242.

[Ch 3]   -   , On partitions of graphs into trees, Dis. Math. 23 (1978), 23-30.

[Ch 4]   -   , On the decomposition of graphs, SIAM J. on Alg. and Dis. Methods 2 (1981), 1-12.

[Ch-Er-Gr 1]   F. R. K. Chung, P. Erdős and R. L. Graham, Minimal de-compositions of graphs into mutually isomorphic subgraphs, Combinatorica 1 (1981), 13-24.

[Ch-Er-Gr 2] F. R. K. Chung, P. Erdös and R. L. Graham, Minimal decompositions of hypergraphs into mutually isomorphic subhypergraphs, (to appear).

[Ch-Er-Gr-Ul-Ya] F.R.K.Chung, P. Erdös, R. L. Graham, S. M. Ulam and F. F. Yao, Minimal decompositions of two graphs into pairwise isomorphic subgraphs, Proceedings of the Tenth South-eastern Conference on Combinatorics, Graph Theory and Computing (1979), 3-18.

[Ch-Er-Sp] F. R. K. Chung, P. Erdös and J. Spencer, On the decomposition of graphs into complete bipartite subgraphs, (to appear).

[Co-Har] E. Cockayne and B. Hartnell, Edge partitions of multipartite graphs, J. Combinatorial Theory (B), 23 (1977), 174-183.

[Do-Ro] J. Doyen and A. Rosa, An extended bibliography and survey of Steiner systems, Proceedings of the Seventh Manitoba Conference on Numerical Mathematics and Computing (1977) Congressus Numerantium XX, 297-361.

[Er] P. Erdös, Problems and results in graph theory and combinatorial analysis, Proc. of the Fifth British Combinatorial Conference, 169-192 Congressus Numerantium, No. XV, Utilitas Math., Winnipeg, Manitoba 1976.

[Er-Fa-Ro-Sc] P. Erdös, R. J. Faudree, C. C. Rousseau and R. H. Schelp, The size Ramsey number, Per. Math. Hung., 9 (1978), 145-161.

[Er-Go-Po] P. Erdös, A. W. Goodman and L. Posa, The representation of graphs by set intersections, Canad. J. Math., 18 (1966), 106-112.

[Er-Gr] P. Erdös and R. L. Graham, On partition theorems for finite graphs, Colloq. Math. Soc. János Bolyai 10, Infinite and Finite Sets, (1973), 515-527.

[Er-Sp] P. Erdös and J. Spencer, Probabilistic Methods in Combinatorics Academic Press, New York 1974.

[Er-Wi] P. Erdös and R. J. Wilson, On the chromatic index of almost all graphs, J. Comb. Theory (B), 23 (1977), 255-257.

[Fi-Wi] S. Fiorini and R. J. Wilson, Edge-colorings of Graphs, Research Notes in Mathematics 16, Pitman, London, 1977.

[For] M. F. Foregger, Hamiltonian decomposition of products of cycles, Dis. Math., 24 (1978), 251-260.

[For-For] M. F. Foregger and T. H. Foregger, the tree covering number of a graph, preprint.

[Ga] C. L. Gabel, A survey of graph decompositions, M.A. thesis, The Pennsylvania State University, 1980.

[Gar-Jo] M. R. Garey and D. S. Johnson, Computers and Intractability, Freeman, San Francisco, 1979.

[Ge] A. Germa, Decomposition of the edges of a complete t-uniform directed hypergraph, dans Combinatorics, Coll. Math. Soc. János Bolyai 18, Keszthely 1976, North Holland (1978), 135-149.

[Gr] R.L. Graham, Rudiments of Ramsey Theory, CBHS Regional Conference Series in Mathematics, no.45, Amer. Math. Soc., Providence, 1981.

[Gr-Li-L] R. L. Graham, S.-Y. R. Li and W.-C W. Li, On the structure of t-designs, SIAM J. on Algebraic and Dis. Methods, 1 (1980), 8-14.

[Gr-Pol] R. L. Graham and H. O. Pollak, On the addressing problem for loop switching, Bell Sys. Tech. Jour., 50 (1971), 2495-2519.

[Gr-Rob] R. L. Graham and R. W. Robinson (to appear).

[Gr-Rot-Sp] R. L. Graham, B. L. Rothschild and J. H. Spencer, Ramsey Theory, John Wiley and Sons, New York, 1980.

[Gr-Sl 1] R. L. Graham and N. J. A. Sloane, On additive bases and harmonious graphs, in SIAM J. on Alg. and Dis. Methods, 1 (1980), 382-404.

[Gr-Sl 2]   -   , Lower bounds for constant weight codes, IEEE Trans. Information Theory, IT-26 (1980), 37-43.

[Gu] R. K. Guy, Unsolved Combinatorial Problems, Combinatorial Mathematics and its Applications, in Proc. Confer. Oxford 1969 (Academic Press, New York 1971), 121-127.

[Gy-Le] A. Gyárfas and J. Lehel, Packing trees of different order into K_n, Proc. Fifth Hungarian Colloq., Keszthely, 1976, Vol. 1, 463-469, Colloq. Math. Soc. János Bolyai, 18, North Holland, Amsterdam, 1978.

[Hal] M. Hall, Jr., Combinatorial Theory, Blaisdell Publishing Co., . Waltham, Mass. 1967.

[Han 1] H. Hanani, On resolvable balanced incomplete block designs, J. Comb. Theory (A), 17 (1974), 275-289.

[Han 2]   -   , Balanced incomplete block designs and related designs, Dis. Math., 11 (1975), 255-369.

[Han-Ray-Wi] H. Hanani, D. K. Ray-Chaudhuri and R. M. Wilson, On resolvable designs, Dis. Math., 3 (1972), 343-357.

[Har] F. Harary, Graph Theory, Addison-Wesley, Reading, Mass. 1969.

[Har-Hs-Mi] F. Harary, D. Hsu and Z. Miller, The biparticity of a graph, J. of Graph Theory, 1 (1977), 131-134.

[Har-Rob-Wo 1] F. Harary, R. W. Robinson and N. C. Wormald, Isomorphic Factorizations, I: Complete graphs, Trans. Amer. Math. Soc., 242 (1978), 243-260.

[Har-Rob-Wo 2]   -   , Isomorphic Factorizations III: Complete multipartite graphs, (to appear).

[Har-Rob-Wo 3]   -   , Isomorphic factorizations V: Directed graphs, Mathematika, 25 (1978), 279-285.

[He-Ko-Ro] P. Hell, A. Kotzig and A. Rosa, Some results on the Ober- wolfach problem, Aequationes Mathematicae, 12 (1975), 1-5.

[He-Ro] P. Hell and A. Rosa, Graph decompositions, handcuffed prisoners and balanced P-designs, Dis. Math., 2 (1972), 229-252.

[Hi] N. Hindman, Large sets and a conjecture of Erdös, Faber and Lovász, preprint.

[Ho] A. J. Hoffman, Eigenvalues and partitionings of the edges of a graph, Lin. Algebra and its Applications, 5 (1972), 137-146.

[Hol] I. Holyer (personal communication).

[Hu] C. Huang, On the existence of balanced bipartite designs II, Dis. Math., 9 (1974), 147-159.

[Hu-Ko-Ro] C. Huang, A. Kotzig and A. Rosa, On a variation of the Oberwolfach problem, Dis. Math., 27 (1979), 261-277.

[Hu-Ko-Ro 2] C. Huang, A. Kotzig and A. Rosa, Further results on tree labellings, preprint.

[Hu-Ro 1] C. Huang and A. Rosa, On the existence of balanced bipartite designs, Utilitas Math., 4 (1973), 55-75.

[Hu-Ro 2]   -   , Decomposition of complete graphs into trees, Ars Combinatoria, 5 (1978), 23-63.

[J 1] Bill Jackson, Edge-disjoint Hamilton cycles in regular graphs of large degree, J. London Math. Soc., 19 (1979), 13-16.

[J 2] Bill Jackson, Decompositions of graphs into cycles (preprint).

[J 3] Brad Jackson, Decompositions of K_{rn} by C_r,(preprint.)

[J 4] -   , Cycle decompositions of K_{2nr+1} by C_r,(preprint.)

[Ka] S. Kageyama, A survey of resolvable solutions of BIBD, Int. Stat. Rev., 40 (1972), 269-273.

[Ko 1] A. Kotzig, On the decomposition of the complete graph into 4k-gons, Mat. Fyz. Časopis, 15 (1965), 229-233.

[Ko 2] -   , Every cartesian product of two circuits is decomposable into two hamiltonian circuits, Centre de Recherche Math., Montréal (1973).

[Ko-Ro] A. Kotzig and A. Rosa, Decomposition of complete graphs into isomorphic factors with a given diameter, Bull. London Math. Soc., 7 (1975), 51-57.

[Lin] S. Lin, Computer solutions of the travelling salesman problem, Bell Syst. Tech. J., 44 (1965), 2245-2269.

[Lo] L. Lovász, On coverings of graphs, in Theory of Graphs, Proc. Coll. Tihany 1966, Academic Press, New York (1968), 231-236.

[Lu] E. M. Luks, Isomorphism of graphs of bounded valence can be tested in polynomial time, 21st Annal Symposium on Foundations of Computer Sciences (1980), 42-49.

[Nash 1] C. St. J. A. Nash-Williams, Decomposition of finite graphs into forests, J. London Math. Soc., 39 (1964), 12.

[Nash 2] -   , Problem in Combinatorial Theory and Its Applications III, North Holland, Amsterdam (1970), 1177-1181.

[Nash 3] -   , Hamilton lines in graphs whose vertices have sufficiently large degrees, in Combinatorial Theory and Its Applications, North Holland, Amsterdam (1970), 813-819.

[Ne-Rod] J. Nešetřil and V. Rödl, Partition theory and its applications, in Surveys in Combinatorics, B. Bollobás, ed., London Math. Soc. Lecture Note Series 38, Cambridge Univ. Press, 1979, 96-156.

[Pa] D. Palumbiny, On decompositions of complete graphs into factors with equal diameters, Boll. Unione Mat. Ital., 7 (1973), 420-428.

[Par] T. Parsons, Ramsey Graph Theory, in Selected Topics in Graph Theory (L. W. Beineke and R. J. Wilson, eds.), Acad. Press, London 1978, 361-384.

[Pe] B. Peroche, On partitions of graphs into linear forests and dissections, preprint.

[Ra-Wi 1] D. K. Ray-Chaudhuri and R. M. Wilson, Solution of Kirkman's Schoolgirl Problem, in Proc. Sym. Pure Math. 19, Amer. Math. Soc. (1971), 187-204.

[Ra-Wi 2] -   , The existence of resolvable designs, in A Survey of Combinatorial Theory (ed. J. N. Srivastava et al) North Holland, American Elsevier, Amsterdam, New York (1973), 361-376.

[Ra-Wi 3] -   , On t-designs, Osaka J. Math., 12 (1975), 737-744.

[Re] A. Recski, Some remarks on the arboricity of the tree-complements, Proc. Faculty of Science, Tokai Univ.; 15 (1979), 71-74.

[Ri] G. Ringel, Problem 25, Theory of Graphs and its Applications, Proc. Sym. Smolenice (1963), Prague, 1974, p. 162.

123

[Ro 1]   A. Rosa, On the cyclic decompositions of the complete
graphs into (4m+2)-gons, Mat. Fyz. Casopis, 16 (1966), 349-353.

[Ro 2]   -   , On certain valuations of the vertices of a graph,
in Theory of Graphs, (ed. P. Rosenthiel), Proc. Symp. Rome,
Dunod, Paris (1967), 349-355.

[Ry]   H. J. Ryser, Combinatorial Mathematics, John Wiley and Sons,
New York, 1965.

[Sl]   N. J. A. Sloane, Hamiltonian cycles in a graph of degree 4,
J. of Comb. Theory, 6 (1969), 311-312.

[So 1]   D. Sotteau, Decompositions de graphes et hypergraphes,
Doctoral dissertation, L'Universite de Paris-Sud, 1980.

[So 2]   -   , Decomposition of K_n^* into circuits of odd length,
Dis. Math., 15 (1976), 185-191.

[So 3]   -   , Decomposition of $K_{m,n}$ into circuits of length
2k, J. Combinatorial Theory (B), (to appear).

[St]   H. J. Straight, Packing trees of different size into the
complete graph, Annals of N.Y. Acad. Sci., 328 (1979), 190-192.

[To]   E. Tomová, On the decomposition of the complete directed graph
into factors with given diameters, Mat. Casopis, 20 91970),
257-261.

[Tu]   W. T. Tutte, On Hamiltonian Circuits, J. London Math. Soc., 21
(1946), 98-101.

[Wa]   W. D. Wallis, Which isomorphic factorizations of regular
graphs are block designs?, J. of Combinatorics, Information and
System Sciences, 2 (1977), 104-106.

[Wi 1]   R. W. Wilson, The necessary conditions for t-designs are
sufficient for something, Utilitas Math., 4 (1973), 207-215.

[Wi 2]   -   , Construction and uses of pairwise balanced
designs, Combinatorics Part I, (eds. M. Hall and J. M. van Lint),
Math. Centre Tracts 55, Math. Centrum Amsterdam (1974), 18-41.

[Wi 3]   -   , An existence theory for pairwise balanced
designs I, II, J. Combinatorial Theory, 13 (1972), 220-273, III.
J. Combinatorial Theory, 18 (1975), 71-79.

[Wi 4]   -   , Decompositions of complete graphs into sub-
graphs, Proc. of the Fifth British Combinatorial Conference,
Aberdeen, 1975, Congressus Numerantium XV, Utilitas Math.
Winnipeg (1976), 647-659.

[Ya]   F. F. Yao, Graph 2-isomorphism is NP-complete, Information
Proc. Letters, 9 (1979), 68-72.

[Za-Lui]   S. Zaks and C. L. Liu, Decomposition of graphs into trees,
Proc. Eighth Southeast Conf. on Combinatorics, Graph Theory and
Computing, 643-654.

[Zn 1]   Š. Znám, Decompositions of the complete directed graph into
factors with given diameters, Comb. Structures and their applications, Proc. Calgary International Conf. (1969), 489-490, Gordan
and Breach, New York, 1970.

[Zn 2]   -   , Decomposition of the complete directed graph into
two factors with given diameter, Mat. Casopis, 20 91970), 254-256.

THE GEOMETRY OF PLANAR GRAPHS

by Branko Grünbaum and G. C. Shephard

Introduction

Every undergraduate course in graph theory mentions basic results about finite planar graphs - Kuratowski's criterion for embeddability, Euler's Theorem, and so on. However the corresponding results for infinite graphs seem to be little known. It turns out that the concept of embeddability in the plane has many ramifications and variants in the infinite case, and one of the purposes of this exposition is to survey these. For the most part results will only be quoted and no proofs given - for these the reader is referred to the literature listed in the bibliography.

In this survey we aim to show how fruitful is the interaction between the theories of finite and of infinite planar graphs. Results from one of these fields often inspire nontrivial problems in the other, and frequently suggest analogous questions about graphs embeddable in manifolds of arbitrary genus.

Although it may seem foreign to the subject, especially to those only interested in problems of a strictly combinatorial or topological nature, quite a large part of what we shall do will be metrical in character. There are several reasons for this. For example, in our discussion of Euler's Theorem and its variants in Section 3, the results are not true unless we impose quite strong restrictions on the kinds of graph we are considering - and we only know how to formulate these restrictions in metrical terms. A second reason is that since we are working in the plane (usually the familiar Euclidean plane), it seems only reasonable to make full use of the large body of known geometrical properties of this space. Finally, there is the personal reason that we became interested in infinite planar graphs because they occur as edge-graphs of tilings or tessellations - and the study of these necessarily has a large geometrical ingredient. On the other hand it is quite possible that the 'correct' conditions for the validity of the results presented can be formulated in graph-theoretic terms. We shall mention some speculations on this topic in Section 3.

The paper is arranged as follows. In the first section, in addition to introducing the necessary terminology, we consider various 'reasonable conditions' that can be imposed on planar

graphs to ensure that, in some sense, they are 'well-behaved'.
Each of these conditions arises naturally in some mathematical
context or other, and a discussion of its consequences can be found
in the literature.   In the second section we shall survey known
conditions for a graph to be embeddable in the plane in such a way
as to have the various properties mentioned in the first section.
Section 3 will be devoted to finding an analogue of Euler's Theorem
that holds for infinite graphs and to some related results (such as
the theorems of Kotzig and Eberhard).   The final section is
concerned with the many problems that arise from considerations of
symmetry, transitivity, and so on.

Throughout we particularly stress unsolved problems and
possible directions for future investigations.

1.   Definitions and conditions

A graph $\Gamma(V, E)$ with vertex set V and edge set E is called a
plane graph if all its vertices and edges lie in a plane π.   To
avoid tiresome special cases and to simplify the exposition we shall
assume (unless otherwise stated) that

(i) Γ is without loops, multiple edges and vertices of valence
2, and

(ii) the vertices of Γ are distinct points, and the edges of
Γ are Jordan arcs of finite length, pairwise disjoint except possibly
at their end-points (vertices).

By π we mean the familiar Euclidean plane with the usual
metric.   Occasionally, however, we find it convenient to suppose
that π can be compactified by adjoining *one* point 'at infinity'.
Such a supposition arises naturally when we wish to relate the plane
π to a 2-sphere by stereographic projection from a point.

An abstract graph Γ is said to be *planar, have a plane
representation,* or *be embeddable in the plane* if it is isomorphic to
a plane graph as defined above.   The plane graph is called a *plane
representation* of Γ.   We shall be careful to distinguish between
the meanings of 'plane' and 'planar'.   We shall also use familiar
terms from graph theory, such as 'finite', 'infinite', 'connected',
'valence', 'circuit', 'tree', and from the theory of polytopes such
as 'face', 'edge-graph', and so on without any further explanation.

For any plane graph Γ, a *region* (*face* or *country*) is defined
as a connected component of the complement of Γ in π.   The richness
of the theory of plane graphs seems to arise, in part, from the
existence of three sorts of 'elements', namely vertices, edges and
regions.   Regions are either bounded or unbounded, and in the case

of finite plane graphs precisely one of the latter sort must occur. If a region is bounded then its closure will be called a *tile*, and if every region of Γ is bounded then the set of all tiles cover the plane without gaps or overlaps.   In this case the set of tiles is called a *tiling* T, and Γ will be referred to as the *edge-graph* of T.   We shall also refer to the edges and vertices of a tile T - these are the elements of Γ that lie on the boundary of T.   In cases which interest us they will clearly form a non-separating circuit in Γ.

Although our principal interest is in infinite graphs, it is frequently convenient to impose various 'finitary' conditions on them.   Examples are the following:

F1   A graph is *finite-valent* if the valence of every vertex is finite.   This is not meant to imply that there is a common upper bound on the valences of all the vertices.   If such an upper bound exists then the graph will be called *uniformly finite-valent*.

F2   A plane graph is called *vertex-accumulation free* if every compact set in the plane contains only a finite number of vertices.

F3   A plane graph is *locally finite* if every bounded set meets only a finite number of edges and vertices (and therefore regions) of the graph.

Condition F3 is the strongest of the above conditions in that it implies both F1 and F2 - it can also be formulated in a uniform manner, but then needs a metrical (in contrast to a topological) definition: A plane graph is *uniformly locally finite* if, for every R > 0, every plane set of diameter R meets at most N elements of the graph - where N depends only on R and not on the particular set of diameter R that is chosen.

In general we shall restrict ourselves to graphs that satisfy F1, F2 or F3.   Further natural restrictions arise from the convexity character of the elements:

C1   A plane graph is *straight* if all its edges are straight line segments.   (This terminology follows Fáry [1948], Tutte [1960], [1963];   unfortunately Thomassen [1977], [1980] uses terms like 'straight line representation' although no straight *lines* are involved.)

C2   A straight plane graph is *convex* if all its bounded regions (or tiles) are convex plane sets (polygons) and its unbounded regions are either convex sets or complements of convex sets.

C3   A convex graph is a *triangulation* if each of its regions (both bounded and unbounded) is a triangle, that is, has three edges and three vertices on its boundary.

To every 3-connected planar graph Γ corresponds a *dual graph* Γ* in which the vertices, edges and regions of Γ* are in one-to-one correspondence with the regions, edges and vertices of Γ, respectively; moreover this correspondence must be inclusion-reversing.   It is easy to see that Γ* is also planar, and if a plane embedding of Γ satisfies condition F3 (local finiteness) then Γ* can be embedded in the plane in such a way as to satisfy F3. In particular, if Γ' is a locally finite plane embedding of Γ, then one can find a locally finite plane embedding Γ*' of Γ* which is *dually situated* in the sense that each vertex of Γ' lies in the interior of the corresponding region of Γ*', each vertex of Γ*' lies in the interior of the corresponding region of Γ', and no two edges intersect unless they correspond to one another and in any case never intersect in more than one point (see Figure 1).   Duality provides another set of criteria which it is convenient to use in classifying plane graphs:

D1   A straight plane graph Γ has the *dual-situation property* if there exists a dually situated dual graph Γ* which is also straight (see Figure 2).

D2   A straight plane graph has the *perpendicular dual-situation property* if it has the dual-situation property and, moreover, every pair of edges that intersect in some point meet at right-angles.

Figure 1 Two plane graphs (one indicated by solid lines and the other by dashed lines) which are dual and dually situated.

128

Figure 2 An example of a finite graph for which condition D1 does not
hold.   There is no straight graph which is dual to this graph and which
is dually situated with respect to it.

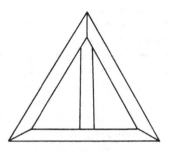

Any plane graph which satisfies D2 is necessarily convex.
In discussing duality properties such as D1 and D2 it is convenient
to slightly modify our definition of 'plane graph' by allowing the
possibility of a 'vertex at infinity'.   In effect we compactify
the plane by adjoining a single point, as mentioned above, and
allow this point to be taken as a vertex.   Of course in this case
we must also allow edges incident with this 'fictitious' vertex to
be unbounded.   Thus in the case of a straight graph every edge
will be either a straight line segment or a *ray*, that is, a half-
line issuing from a vertex, and all such rays point away from a
single point in the plane (see Figure 3).   If R represents the
infinite region of a finite plane convex graph Γ, then it is
usually convenient to take the vertex of the dual graph Γ* that
corresponds to R to be the fictitious vertex.
       Properties related to duality frequently stem from the
relationship between graphs and polytopes, polyhedra and similar
sets.   We briefly review the necessary definitions.
       A *polyhedron* (in 3-dimensional Euclidean space) is any set
which can be written as the intersection of a finite number of
closed half-spaces.   A polyhedron which is bounded is called a
polytope or *3-polytope*.   Polyhedra and polytopes are necessarily
convex sets.   An *apeirohedron* (in 3-dimensional Euclidean space)
is any convex set whose boundary is the locally finite union of
infinitely many convex polygons or unbounded convex regions bounded
by locally finite unions of line segments.   An *apeirotope* is an
apeirohedron all of whose faces are convex polygons.
       P1 A graph is called *3-polytopal* if it is isomorphic to the

Figure 3 A graph (solid lines) with a 'vertex at infinity'. The rays
all point away from the dot near the centre of the diagram. This graph
has property D2 - a dual graph with the perpendicular dual-situation
property is indicated by dashed lines.

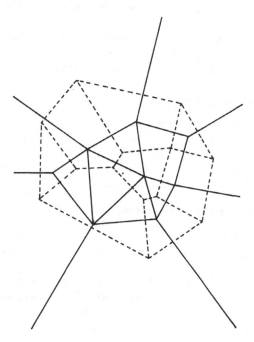

edge-graph of a polytope.

Certainly a polytopal graph Γ is finite and planar.   The
latter follows from the fact that a plane embedding of Γ can be
constructed by projecting the corresponding polytope Π stereo-
graphically onto the plane π from a suitable point.   In this case
the image of Π is often called a *Schlegel diagram* for Π.   It does
not follow, of course, that any polytopal plane graph is necessarily
the Schlegel diagram of some polytope.   In fact, convenient
criteria for recognising when a polytopal plane graph *is* a Schlegel
diagram are unknown, and much has been written about the
corresponding problem in higher dimensions.

P2   A graph is *apeirohedral* if it is isomorphic to the edge-
graph of an apeirohedron.

P3   A graph is *apeirotopal* if it is isomorphic to the edge-graph of an apeirotope.

An apeirotopal graph is infinite and planar.   By comparison with polytopal graphs, little seems to be known about this kind of graph;   references to the relevant literature will be given in the next section.

We postpone to Section 4 the discussion of other restrictions that it is sometimes convenient to place on a plane graph related to symmetries, groups of isomorphisms, transitivity, and so on.

2.   Criteria for plane embeddability

We now state a number of problems concerning the possibility of embedding an abstract graph Γ in the plane.   In many cases we shall be interested in those graphs of which a plane representation Γ' satisfies one or more of the additional conditions stated in the previous section.   The first problem is fundamental:

Characterise graphs which are planar.

For finite graphs the original solution to this problem was given by the well-known theorem of Kuratowski [1930].   Variants of this characterisation are given by Whitney [1933], Mac Lane [1937], Tutte [1958] and others.   For infinite graphs character-isations have been given by Dirac & Schuster [1954], Wagner [1967] and others.   Halin [1966] and R. Schmidt [1980] characterised those graphs Γ for which there exists a plane representation Γ' satisfying condition F2 (vertex-accumulation free).

Characterise planar graphs Γ which admit a straight plane representation Γ'.

The solution to this problem is given by the following.

Theorem 1   *Every planar graph Γ has a plane representation Γ' which is straight (condition C1).*

The finite case of this result was formulated and proved by Wagner [1936] and Fáry [1948].   With regrettable lack of fairness to Wagner, many authors continue to call this result 'Fáry's Theorem'.   A proof of Theorem 1 also follows easily from the theorems of Steinitz [1916], Stein [1951] and others, discussed below.

In the infinite case, Theorem 1 is due to Thomassen [1977] who showed, moreover, that the straight plane representation Γ'

could be chosen so as to be contained in a bounded set. He also
established, amongst others, the following result:

Theorem 2 *Every plane graph satisfying conditions F1 and F2
(that is, is finite-valent and vertex-accumulation free) is
isomorphic to a subgraph of a triangulated plane graph (condition C3).*

The finite version of Theorem 2 is due to Wagner [1936]. In
reading Thomassen's version of his proof one should take note that
he uses 'locally finite' to mean that the graph satisfies F1
(finite-valent) instead of the more usual meaning adopted here.

Tutte [1936] and Thomassen [1980], [1980*] consider
simultaneous straight embeddings of a graph Γ and its dual Γ^*.
They show that if Γ is a 3-connected planar graph, then such
embeddings are possible that have the dual-situation property D1.
In the case when Γ is finite, the dual Γ^* will have a 'vertex at
infinity' as explained in the previous section. It is an open
problem whether it is also possible to always choose the plane
representations of Γ and Γ^* so that the stronger condition D2
(perpendicular dual-situation) is satisfied.

It is extremely easy to see that *every* graph Γ has a straight
representation in 3-dimensional Euclidean space (Wagner [1966]).
For example one need only choose the vertices of the representation
as distinct points on the moment curve $\{(t, t^2, t^3) \mid -\infty < t < \infty\}$.

We now make some observations and raise a number of problems
concerning the analogous results for compact manifolds. It is
classical that every compact 2-manifold can be metrised in such a
way that it has constant curvature. A graph embedded in a
2-manifold metrised in this way may be called *straight* provided
each of its edges is represented by a geodesic arc. It is well-
known (and it also follows from a variety of results such as those
discussed in Section 4) that every finite graph embeddable in the
sphere (or in the projective plane) admits a straight representation.
We conjecture that the situation is the same for infinite graphs,
and for (finite or infinite) graphs embeddable in other compact
manifolds, orientable or not.

*Characterise planar graphs which have convex plane
representations.*

Although a complete solution of this problem is probably not
hard in view of known results, it does not seem to be available in
the literature. Among the published results, the simplest to
formulate is the following, proofs of which (in some cases in

stronger versions) were given by Stein [1951], Tutte [1960], [1963], Thomassen [1980]; it is also a corollary of the theorem of Steinitz [1916] which is discussed below, and, in turn, implies the result of Wagner [1936] about straight triangulations quoted above (the finite case of Theorem 2).

Theorem 3 *Every finite 3-connected planar graph Γ has plane representations that are convex (property C2).*

It would be of interest to find analogues of Theorem 3 for graphs embeddable in a compact 2-manifold M. (In such a manifold M a set S is called *convex* if for any two points x, y of S, a geodesic arc of M with end-points x, y is contained in S.) For the sphere the result can be deduced from Steinitz's Theorem, while a result for the case of the torus follows from Theorem 13.

Various strengthenings of Theorem 3 are known. For example, any non-separating circuit C in Γ can be chosen as the boundary of the unbounded face; if C is a circuit of n edges, any convex n-gon can be chosen to represent C, and so on. However the following result of Thomassen [1980] is by far the most important, since not only does it include all these statements as corollaries, but can also be used to prove many results on infinite graphs (see below). As the number of applications is likely to increase we present Thomassen's result here, in spite of the fact that its formulation is rather technical.

Theorem 4 *Let S be a non-separating circuit in a 2-connected finite plane graph Γ, and let P_1, P_2,..., P_k be edge-disjoint paths in Γ with union S. Let C be the boundary of any convex k-gon in the plane and let S be mapped homeomorphically onto C in such a way that each side of C is the image of one of the paths P_i. Then the mapping of S onto C can be extended to a convex representation of Γ with S as the boundary of the unbounded region, if and only if the following conditions hold:*

(i) Each vertex V of Γ which has valence at least 3, and is not contained in S, is joined to S by three paths which are pairwise disjoint except for V.

(ii) No S-component of Γ has all its vertices of attachment in a single P_i. (An S-component of Γ is defined as a connected component of the complement of S in the set of edges and vertices of Γ.)

(iii) Each circuit of Γ, which shares no edges with S, has at least three vertices of valence 3.

An example illustrating Theorem 4 is given in Figure 4. To illustrate the strength of Thomassen's result we mention one of its corollaries; another which relates to double-periodic graphs will be given in Section 4.

Theorem 5 *Every plane graph Γ which is 3-connected and has properties F1 (valence-finite) and F2 (vertex-accumulation free) has a convex representation.*

This theorem was proved by Thomassen using a result on convex embeddings of trees with no vertex of valence 1, and Theorem 4. A strengthened version of this result on trees will be presented in Theorem 8 below.

It is not hard to strengthen Theorems 3 and 5 by assuming that the original graphs are plane graphs and then insisting that the required convex representations can be obtained through continuous deformations (isotopies). Deeper is the following result of Cairns [1944], [1944*] (and for extensions see Bing & Starbird [1978], [1978*], Ho [1973], [1975]).

Theorem 6 *If Γ_0 and Γ_1 are plane graphs with property C3 (that is, triangulations), and if there exists an orientation-preserving homeomorphism of the plane that maps Γ_0 onto Γ_1, then Γ_0 and Γ_1 are linearly isotopic.*

Figure 4 This figure illustrates Thomassen's Theorem (Theorem 4). The paths P_1, P_2, P_3 and P_4 are taken as the arcs X_1X_2, X_2X_3, X_3X_4, X_4X_1 making up the boundary of the unbounded region of the graph Γ shown in (a). If the edge E is deleted then condition (i) of the theorem is violated for vertex V; condition (ii) is violated for the S-component of V; condition (iii) is violated for the circuit of length three containing W. Part (b) shows a convex embedding of Γ with P_1, P_2, P_3, P_4 mapping into the sides of a square.

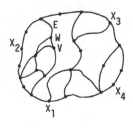
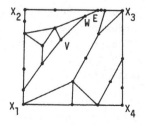

By 'linearly isotopic' we mean that there exists a family
$\{\Gamma(t)|\ 0 \le t \le 1\}$ of straight plane graphs which depend
continuously on the parameter t, each $\Gamma(t)$ is isomorphic to Γ_0
(and Γ_1), and $\Gamma(0) = \Gamma_0$, $\Gamma(1) = \Gamma_1$.

Probably Theorem 6 is still true if Γ_0 and Γ_1 are straight
graphs (and not necessarily triangulations) as well as for infinite
graphs. Also if Γ_0 and Γ_1 are convex we conjecture that it is
possible to choose the family $\{\Gamma(t)|\ 0 \le t \le 1\}$ used in defining the
isotopy as a convex isotopy, that is, so that each of the $\Gamma(t)$ is
convex.

The theorem of Steinitz [1916] mentioned above, that leads
immediately to a proof of Theorem 3, can be stated as follows:

<u>Theorem 7</u> *A finite graph* Γ *is 3-polytopal (condition P1) if and
only if it is planar and 3-connected.*

For detailed proofs of Theorem 7 see Steinitz & Rademacher
[1934], Grünbaum [1967], Barnette & Grünbaum [1969]. In analogy
with Theorem 4 it can be shown that the polytope Π (used in defining
the property P1 in the previous section) can have the shape of one
face arbitrarily prescribed, or an arbitrary circuit in Γ can be
chosen so that the corresponding edge-circuit on Π is a 'shadow-
boundary' of Π (see Barnette & Grünbaum [1970], Barnette [1970]).
However, neither of these results imply Theorem 4, nor are they
implied by it. In Steinitz & Rademacher [1934] it is also
established that, in analogy with Theorem 6, if a graph Γ is
represented as the edge-graph of two different polytopes Π_0 and Π_1,
then these are 'polytopally isotopic' in the sense that there exists
a continuous family of polytopes joining Π_0 to Π_1 (or to the mirror
image of Π_1) and each polytope in this family has edge-graph
isomorphic to Γ.

As generalisations of Theorems 5 and 7 we conjecture that
every finite-valent, vertex-accumulation free 3-connected planar
graph is apeirohedral (satisfies condition P2) and, if all its faces
are bounded, is apeirotopal (satisfies condition P3). Unfortunately
the attempt to prove these conjectures in a way analogous to
Thomassen's proof of Theorem 5 fails since there is no suitable
analogue of Theorem 4. However, the other ingredient of Thomassen's
proof (Theorem 8.4 of Thomassen [1980]) can be strengthened to the
following easily proved result:

<u>Theorem 8</u> *Every finite-valent tree with no vertices of valence 1
or 2 is apeirohedral (condition P2).*

We remark that (extending the comment following Theorem 2 above) one can show that every graph Γ is isomorphic to a subgraph of the edge-graph of a 4-polytope Π (Carathéodory [1907]). To see this one needs only choose representatives of the vertices of Γ as points on the moment curve $\{(t, t^2, t^3, t^4)\mid -\infty < t < \infty\}$ and take their convex hull. However no characterisation of graphs which are isomorphic to the edge-graphs of n-dimensional polytopes, apeirohedra or apeirotopes is known for any $n \geq 4$. In other words, Theorems 7 and 8 appear to have no higher dimensional analogues.

3. Euler's Theorem and its derivatives

In this section we consider the possibility of extending Euler's Theorem to infinite planar graphs. In other words we seek to adapt the usual formulation

$$v - e + t = 2 \qquad\qquad (1)$$

for finite planar graphs to the infinite case. Here v, e and t represent the numbers of vertices, edges and regions (both bounded and unbounded) in any plane representation of a connected planar graph. Since, in the case of infinite graphs, by definition, the values of v, e and t are all infinite, it is clear that any suitable reformulation of (1) must involve some sort of limiting process, and it turns out that a proper discussion of such a process is rather delicate.

It is perhaps worthwhile to record the fact that the extension of Euler's Theorem to infinite graphs is probably the branch of mathematics about which most nonsense has been written and published! One author after another has assumed that any possible interpretation they could put on the equation (1) in the infinite case *must* be true for *all* graphs, and have sought to show that this is so by specious arguments such as drawing graphs on 2-spheres 'and letting the radius of the sphere expand to infinity'! And such 'arguments' are not confined to an earlier 'non-rigorous' phase of mathematics, but examples could be quoted of such publications in the last few years.

To see the sort of problems that arise, we may consider a tiling T such as that in Figure 5. Only a small portion of the tiling can be shown in the diagram, but it should be clear how the tiling can be continued indefinitely in every direction. The tiles are triangles and the valence of each vertex is seven. If we divide (1) by t, let $t \to \infty$, and suppose that the limits

$$V = \lim_{t \to \infty} \frac{v}{t}, \qquad\qquad E = \lim_{t \to \infty} \frac{e}{t}, \qquad\qquad (2)$$

exist (in some sense), then we obtain

$$V - E + 1 = 0. \qquad\qquad (3)$$

The numbers V and E have the obvious interpretations as the 'number of vertices per tile' and 'the number of edges per tile' respectively. Now, from the diagram we see that each tile has three edges each of which is shared by one other tile, so E must be 3/2; and each tile has three vertices each of which is shared by seven tiles, so V must be 3/7. However these values do not satisfy (3); this leads to the conclusion that such a superficial approach to the problem is completely spurious.

Moreover the difficulty is not overcome by imposing conditions on the tiling such as those described in Section 1. The tiling T of Figure 5 is locally finite (satisfies F3 and therefore all the finitary conditions), is a triangulation (satisfies C3 and therefore all the convexity conditions), and has the dual-situation property (condition D1). Moreover it is not difficult to see how it can be slightly modified so as to have the perpendicular dual-situation property (D2) also. Finally, it is apeirotopal (condition P3).

It seems as if a sufficient condition on a tiling for an analogue of Euler's Theorem to hold is necessarily metrical. We shall begin by introducing some terminology.

Let us say that a tiling (or graph) is *normal* if its tiles (regions) are uniformly bounded, that is, there exist two parameters u and U such that every tile (region) contains a circular disk of radius u and is contained in a circular disk of radius U. (We recall our original assumptions about graphs: they contain no loops, multiple edges, or vertices of valence two. It follows easily that in any tiling each tile must be a closed topological disk with at

Figure 5 A non-normal tiling.

least three edges and three vertices on its boundary. The uniform boundedness condition just described ensures that the tiling is *uniformly* locally finite, *uniformly* finite-valent, and that the number of tiles neighbouring on any tile is also uniformly bounded. The tiling of Figure 5 is not normal.) Intuitively we may regard the condition of normality as implying that none of the tiles is very long and thin.

One of the important properties of a normal tiling is given by the following lemma, which is used in the proof of many of the assertions that follow. To state it we need to explain the notion of a patch of tiles. Let P be any point of the plane, r > 0 a real number, and D(r,P) be the circular disk centred at P and of radius r. Then we write A(r, P) for the subset of tiles which have non-empty intersection with D(r, P) together with the minimum number of additional tiles needed to make the union of the tiles in A(r, P) simply connected (see Figure 6). The union of the tiles in A(r, P) is a topological disk, and A(r, P) is called a *patch* of tiles (from the tiling T defined by the disk D(r, P)). We write Γ(r, P) for the (finite plane) edge-graph of A(r, P), v(r, P) and e(r, P) for the numbers of vertices and edges in Γ(r, P), and t(r, P) for the number of tiles in A(r, P).

Figure 6 A patch of tiles. This consists of all tiles that intersect the circular disk D, together with further tiles (hatched) that must be adjoined to make the union of the tiles simply connected.

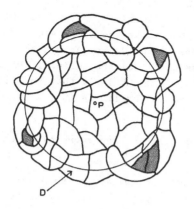

Normality Lemma *In any normal tiling, for any* $x \to 0$ *and any point*
P *in the plane,*

$$\lim_{r\to\infty} \frac{t(r + x, P) - t(r, P)}{t(r, P)} = 0.$$

(The numerator of the expression can be regarded as a measure of
the number of tiles round the outside of the patch, and the lemma
asserts that for large r, this number becomes negligible compared
to the number of tiles in the patch.)   For a proof of the normality
lemma, see Grünbaum & Shephard [1981, Section 3.2].

Euler's Theorem (in its usual form) can be applied to the
graph $\Gamma(r, P)$ to yield

$$v(r, P) - e(r, P) + t(r, P) = 1. \tag{3}$$

(Notice that in evaluating $t(r, P)$ we do not count the unbounded
region.)   Dividing (3) by $t(r, P)$ we obtain

$$\frac{v(r, P)}{t(r, P)} - \frac{e(r, P)}{t(r, P)} + 1 = \frac{1}{t(r, P)}. \tag{4}$$

Now let $r \to \infty$ and suppose that the limits

$$v(T) = \lim_{r\to\infty} \frac{v(r, P)}{t(r, P)} \quad \text{and} \quad e(T) = \lim_{r\to\infty} \frac{e(r, P)}{t(r, P)} \tag{5}$$

exist, and we arrive at

$$v(T) - e(T) + 1 = 0 \tag{6}$$

which is analogous to (3).   We remark that the condition of
normality does *not* imply that the limits (5) exist, but it does
imply that if they exist then they take the same values whatever
point P is chosen as the centre of the disks used in defining the
patches.   Equally, as we have seen from the tiling in Figure 5,
the existence of the limits (5) does not imply that the tiling is
normal.   A normal tiling for which the limits (5) exist is called
a *balanced* tiling.

Theorem 9 Euler's Theorem for Tilings *For every balanced tiling
the relation* (6) *holds.*

Having obtained this result it is natural to examine the
various theorems concerning finite planar graphs which are
consequences of Euler's Theorem, and see if they can be extended to
the infinite case.   A good example is provided by Kotzig's Theorem.

To state this we define the *weight* w(E) of an edge E in a graph
to be the sum of the valences at its two end-points, and the weight
w(Γ) of a graph Γ to be the minimum of the weights of all the edges
in Γ.

Theorem 10   Kotzig's Theorem for Finite Planar Graphs   *For every*
polytopal graph Γ, w(Γ) $\leq$ 13.

The analogous result for infinite graphs (tilings) appears to
be that, under certain additional assumptions, we have w(Γ) $\leq$ 15.
We conjecture that this result is true for all normal tilings, and
we can prove it for 'strongly balanced tilings' (see below).
However, the only published proof (see Grünbaum & Shephard [1981*])
relates to doubly-periodic tilings, for information about which we
refer the reader to Section 4.   Recent work of Zaks [1981] extends
and improves these results and examines Kotzig-type theorems for
graphs embedded in 2-manifolds of positive genus.

In some ways the definition of balance is extremely weak, and
for other extensions of Euler's Theorem additional assumptions are
necessary.   If we use $t_j(r, P)$ and $v_j(r, P)$ to denote the number
of tiles with j edges (and therefore j vertices) and the number of
vertices of valence k, in the patch $A(r, P)$, respectively, then the
tiling T is called *strongly balanced* if all the limits

$$t_j(T) = \lim_{r \to \infty} \frac{t_j(r, P)}{t(r, P)}, \qquad j = 3, 4, \dots$$

and

$$v_k(T) = \lim_{r \to \infty} \frac{v_k(r, P)}{t(r, P)} \qquad k = 3, 4, \dots \tag{7}$$

exist.   In this case we can speak of the 'proportion of tiles in
Γ which have j edges' and the 'number of k-valent vertices per tile'
with justification, for the normality lemma can be used to show that
all these quantities are independent of the choice of P.

One application of these definitions is to the infinite form
of Eberhard's Theorem.   We begin by recalling the result in the
case of finite graphs.

Let us write, for a polytopal graph Γ, e for the number of
edges, t_j for the number of regions in Γ with j sides (including the
infinite region) and v_k for the number of vertices in Γ of valence
k.   Then it is easy to show, using Euler's Theorem, that

$$2e = \sum_j jt_j = \sum_k kv_k,$$

$$\left.\begin{array}{l} \sum_j (4 - j)t_j + \sum_k (4 - k)v_k = 8, \\[2mm] \sum_j (6 - j)t_j + 2\sum_k (3 - k)v_k = 12, \\[2mm] 2\sum_j (3 - j)t_j + \sum_k (6 - k)v_k = 12. \end{array}\right\} \qquad (8)$$

If the graph is trivalent (has all vertices of valence 3) then the second equation in (8) becomes

$$\sum_j (6 - j)t_j = 12. \qquad (9)$$

Here, and in (8), all summations are over all $j \geq 3$ and $k \geq 3$. It will be noticed that the value t_6 does not appear in (9) for its coefficient is zero. Then we have the following:

Theorem 11 Eberhard's Theorem *If t_3, t_4, t_5, t_7, t_8,... are any non-negative integers that satisfy (9), then there exists a polytopal graph for which t_j is the number of j-sided faces for all $j \geq 3$, $j \neq 6$.*

A proof of this result can be found in Grünbaum [1967, Section 13.3].

For a strongly balanced tiling T we can prove the following analogues of (8) using Euler's Theorem for Tilings:

$$\left.\begin{array}{l} \sum_j (4 - j)t_j(T) + \sum_k (4 - k)v_k(T) = 0, \\[2mm] \sum_j (6 - j)t_j(T) + 2\sum_k (3 - k)v_k(T) = 0, \\[2mm] 2\sum_j (3 - j)t_j(T) + \sum_k (6 - k)v_k(T) = 0, \end{array}\right\} \qquad (10)$$

and an analogue of Eberhard's Theorem can be proved (see Grünbaum & Shephard [1981**]):

Theorem 12 Eberhard's Theorem for Tilings *If t_3, t_5, t_6,... and v_3, v_5, v_6,... are non-negative real numbers (of which only a finite number are non-zero) which satisfy*

$$\sum_j (4 - j)t_j + \sum_k (4 - k)v_k = 0,$$

*then there exists a strongly balanced tiling, and a real positive
number γ such that*

$$t_j(T) = \gamma t_j, \qquad v_k(T) = \gamma v_k,$$

for all $j \geq 3$, $j \neq 4$ *and all* $k \geq 3$, $k \neq 4$. *Further, if*
$\sum_j t_j + \sum_k v_k = \kappa \neq 0$, *then* $\gamma > \frac{1}{3}\kappa^{-1}$.

Analogous statements hold corresponding to the other two
equations of (10).

Eberhard's Theorem has been extended in many other directions
as well; for some of these extensions see, for example, Grünbaum &
Zaks [1974], Jucovič [1976], Jendrol & Jucovič [1977], Kraeft [1977],
Enns [1980] and the references given in these papers.

Probably the most interesting question from a graph-theoretic
point of view is whether the notions of normality, balance and
strong balance can be replaced by non-metric graph-theoretic
conditions. In any such attempt, the following type of
construction will probably play an important part. Let P be any
vertex of the (infinite) graph Γ, and k a non-negative integer.
Instead of the patch A(r, P) we consider the induced subgraph
B(k, P) of Γ, which has as vertices all those vertices of Γ which
can be reached from P by paths of (graph-theoretic) length at most
k, together with those additional vertices of G which are separated
from the 'far' vertices by the vertices already included in B(r, P).
Properties analogous to normality seem to depend on the finiteness
of the limit

$$\lim_{k \to \infty} \frac{V(k + 1, P) - V(k, P)}{k}$$

where V(k, P) denotes the number of vertices of the graph B(r, P).

4. Symmetry properties of plane graphs

Associated with any graph Γ is a group G(Γ) of self-
isomorphisms of Γ. If Γ is planar then each plane embedding Γ'
of Γ has a symmetry group S(Γ'), namely the group of isometries
(congruence transformations) which map Γ' onto itself. In a
similar way we can define the symmetry group S(Π) of a 3-polytope Π.
The purpose of this section is to discuss properties of, and
relations between, these groups.

To begin with we shall consider properties relating to
plane tilings which are doubly-periodic, that is to say, tilings
which admit as symmetries translations in at least two non-parallel

directions.   Such a tiling can be thought of as consisting of
infinitely many translations of a small portion of the tiling,
namely that lying within a fundamental parallelogram (see Figure 7).

Theorem 13   *If a locally finite 3-connected plane graph is doubly-
periodic, then it is isomorphic to a doubly-periodic convex plane
graph.*

   This result was first established by Mani-Levitska, Guigas &
Klee [1979] using complicated arguments.   However it can also be
deduced very simply by applying the result of Thomassen [1980]
(given above as Theorem 4) to a suitably chosen 'fundamental region'
of the graph.
   For doubly-periodic 3-connected planar graphs Γ it is not
known whether both Γ and its dual Γ^* can admit doubly-periodic
convex embeddings which are dually situated (condition D1 of Section
1).
   One of the fundamental questions in this area is the following:

   *Given a planar graph Γ, does there always exist a plane
embedding Γ' of Γ such that the groups $G(\Gamma)$ and $S(\Gamma')$ are isomorphic?*

   It is easy to see that in the case of finite graphs the
answer is, in general, negative.   Only in comparatively trivial

Figure 7   A doubly-periodic plane graph with a fundamental parallelogram
indicated by shading; the graph consists of translations of the part of
it that lies within this parallelogram.

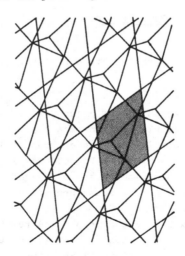

cases can S(Γ') be made equal to G(Γ). However, in some ways the
question is not natural in the finite case, and it is more
appropriate to ask whether, for any polytopal graph Γ, there exists
a polytope Π whose edge graph is isomorphic to Γ and whose symmetry
group S(Π) is isomorphic to G(Γ). The answer to this variant is
in the affirmative; special cases were established by Grünbaum
[1967, Section 13.2] and by Barnette [1971]. The general
affirmative solution is due to Mani [1971]. These arguments can
easily be modified to yield an affirmative answer for all finite
planar graphs and their embeddings in a sphere. It may be
conjectured that the same is true for every compact 2-manifold M
with suitable metric, and every finite graph embeddable in M.

In the case of infinite planar graphs it is easy to see that
in general the answer is also in the negative. A simple example
is provided by the tiling in Figure 5 - no plane embedding of this
graph Γ can have symmetry group isomorphic to G(Γ). It would
therefore seem natural to rephrase the question restricting the kind
of graphs under consideration. A candidate for this restriction
(which will seem more natural after our comments on transitivity)
is to assert that Γ must have a normal representation, but the
problem as to whether there exists a suitable graph-theoretic
condition is open.

However, more promising is the approach in which the
Euclidean metric of the plane is replaced by a suitable (hyperbolic)
metric. In this case it is known that the 'regular' tilings
{p, q}, in which the tiles are p-gons meeting q at each vertex, can
indeed be represented by a suitable metrically-regular convex
tiling in the hyperbolic plane (see, for example, Coxeter & Moser
[1972, Chapter 5], Fejes Tóth [1965, Section 1.3.3]). But even in
this interpretation it is not known whether Γ' can be chosen so that
S(Γ') and G(Γ) are isomorphic (see a discussion of this and related
questions in Grünbaum & Shephard [1979, Section 4]).

A graph Γ is called *homeogonal* or *vertex-transitive* if the
vertices form one transitivity class under the group G(Γ) and is
called *2-homeogonal* or *2-vertex-transitive* if the vertices form two
transitivity classes. In a similar way, terms like *homeotoxal* or
edge-transitive are defined.

We begin by considering homeogonal 3-connected planar graphs.
In the case of finite graphs Γ there are exactly 18 types and two
infinite families of types (each depending on a positive integer
parameter) corresponding to the edge-graphs of the Platonic and
Archimedean solids and the families of prisms and antiprisms.

Figure 8 The eleven types of 3-connected homeogonal (vertex-transitive) normal plane tilings. One of the eleven occurs in two enantiomorphic forms, both of which are shown.

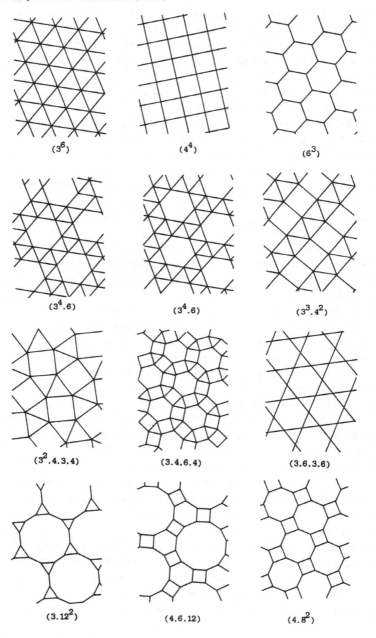

(3^6) (4^4) (6^3)

$(3^4.6)$ $(3^4.6)$ $(3^3.4^2)$

$(3^2.4.3.4)$ $(3.4.6.4)$ $(3.6.3.6)$

(3.12^2) $(4.6.12)$ (4.8^2)

It will be seen that here $G(\Gamma)$ and $S(\Gamma)$ are isomorphic. In the
case of infinite planar graphs, even if *all* the relevant conditions
of Section 1 are imposed, there are infinitely many types;
examples are provided by the 'regular' tilings {p, q} mentioned
above. These exist for all p, q such that $p^{-1} + q^{-1} < \frac{1}{2}$; the case
p = 3, q = 7 is shown in Figure 5. However, if we restrict
attention to graphs which admit a *normal* plane representation, then
the situation changes radically, and we find that there are just 11
different types (see Figure 8). These are known as the Archimedean
tilings (see Grünbaum & Shephard [1981, Section 2.1]). Examining
this figure reveals the remarkable fact that not only is each type
representable by a convex tiling Γ' with $S(\Gamma')$ isomorphic to $G(\Gamma)$,
but every tile is a convex *regular* polygon, and moreover the tiling
is *isogonal* (by which we mean that $S(\Gamma')$ is transitive on the
vertices). All these tilings are well-known and have been
discovered and rediscovered many times. It seems that they were
originally described by Kepler as long ago as 1619.

For finite 3-connected planar tilings there are just nine
types which are homeotoxal; as in the case of vertex-transitive
tilings these can be represented by the edges of polyhedra which
admit the same groups of isometries (see for example, Fleischner &
Imrich [1979], Grünbaum & Shephard [1981***]). For infinite
tilings Γ which admit normal plane embeddings there are exactly
five types, see Figure 9. Similar remarks to the above apply here:
there exists a plane convex representation Γ' such that $S(\Gamma')$ is
isomorphic to $G(\Gamma)$ and moreover Γ' is *isotoxal* (by which we mean
that $S(\Gamma')$ is transitive on the edges of Γ'). For non-normal
tilings Γ the situation is again unresolved. Similarly open are
the questions concerning homeogonal or homeotoxal finite graphs
embeddable in 2-manifolds of higher genus. Corresponding
statements about graphs Γ for which the group $G(\Gamma)$ is transitive
on the regions or tiles can be deduced from the results for
homeogonal graphs by duality.

The problem of determining planar graphs which are 2-homeo-
gonal is much more complicated. In the case of finite planar
graphs it should be possible to determine the number of types
relatively easily, but, so far as we are aware, this has never been
done. In the case of normal graphs which admit a normal plane
representation, and in which every two tiles intersect in a
connected set, we already know of 503 types, but this enumeration
may not be complete. For a discussion of this problem (in the
dual formulation) see Grünbaum, Löckenhoff, Shephard & Temesváry
[1981]. Here again all known types of 2-homeohedral tiling Γ have

convex plane representations Γ' such that S(Γ') and G(Γ') are
isomorphic and Γ' is 2-isohedral (by which we mean that the tiles
form two transitivity classes under the symmetry group S(Γ')).
Whether this holds in general is still an open question.

We hope that the material presented in this paper has shown
that the subject of infinite planar graphs is still rich in open
problems and that this exposition and survey will encourage further
work in this field.

Figure 9 There are five types of 3-connected homeotoxal (edge-transitive)
normal plane tilings. Four of these are the tilings (3^6), (4^4), (6^3) and
(3.6.3.6) shown in Figure 8. The fifth one is shown here.

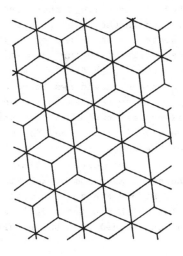

REFERENCES

D. W. Barnette

[1970]   Projections of 3-polytopes.
Israel J. Math. 8(1970), 304-308. .

[1971]   The graphs of polytopes with involutory automorphisms.
Israel J. Math. 9(1971), 290-298.

D. W. Barnette and B. Grünbaum

[1969]   On Steinitz's theorem concerning convex 3-polytopes
and on some properties of planar graphs.
"The Many Facets of Graph Theory"
Lecture Notes in Mathematics Vol. 110, Springer, Berlin
1969, 27-40.

[1970]   Preassigning the shape of a face.
Pacific J. Math. 32(1970), 299-306.

R. H. Bing and M. Starbird

[1978]   Linear isotopies in E^2.
Trans. Amer. Math. Soc. 237(1978), 205-222.

[1978*]   Super triangulations.
Pacific J. Math. 74(1978), 307-325.

S. S. Cairns

[1944]   Isotopic deformations of geodesic complexes on the
2-sphere and the plane.
Ann. of Math. (2) 45(1944), 207-217.

[1944*]   Deformations of plane rectilinear complexes.
Amer. Math. Monthly 51(1944), 247-252.

C. Carathéodory

[1907]   Ueber den Variabilitätsbereich der Koeffizienten von
Potenzreihen, die gegebene Werte nicht annehmen.
Math. Ann. 64(1907), 95-115.

H. S. M. Coxeter and W. O. J. Moser

[1972]   Generators and Relations for Discrete Groups.
(Third Edition)
Springer, Berlin Heidelberg New York 1972.

G. A. Dirac and S. Schuster

[1954]   A theorem of Kuratowski.
Indag. Math. 16(1954), 343-348.

T. C. Enns

[1980]   Convex 4-valent polytopes.
Discrete Math. 30(1980), 227-234.

I. Fáry

[1948]   On straight line representations of planar graphs.
Acta Sci. Math. Szeged 11(1948), 229-233.

L. Fejes Tóth

[1965]   Reguläre Figuren.
Akadémiai Kiadó, Budapest.
English translation: Regular Figures.
Pergamon, New York 1964.

148

H. Fleischner and W. Imrich

[1979] Transitive planar graphs.
Math. Slovaca 29(1979), 97-106.

B. Grünbaum

[1967] Convex Polytopes.
Wiley, London 1967.

B. Grünbaum, H.-D. Löckenhoff, G. C. Shephard and A. H. Temesváry

[1981] The enumeration of 2-tile-transitive tilings.
(In preparation)

B. Grünbaum and G. C. Shephard

[1981] Tilings and Patterns.
Freeman, San Francisco (to appear).

[1981*] Analogues for tilings of Kotzig's Theorem on minimal
weights of edges.
"Theory and Practice of Combinatorics"
Annals of Discrete Mathematics (to appear).

[1981**] The theorems of Euler and Eberhard for tilings of the
plane.
Resultate der Math. (to appear).

[1981***] Spherical tilings with transitivity properties.
"The geometric vein: Essays presented to H. S. M. Coxeter"
(to appear).

B. Grünbaum and J. Zaks

[1974] The existence of certain planar graphs.
Discrete Math. 10(1974), 93-115.

R. Halin

[1966] Zur häufungspunktfreien Darstellung abzählbarer Graphen
in der Ebene.
Arch. Math. 17(1966), 239-243.

C.-W. Ho

[1973] On certain homotopy properties of some spaces of linear
and piecewise linear homeomorphisms. I, II.
Trans. Amer. Math. Soc. 181(1973), 213-233, 235-243.

[1975] Deforming PL homeomorphisms on a convex 2-disk.
Bull. Amer. Math. Soc. 81(1975), 726-728.

S. Jendrol and E. Jucovič

[1977] Generalization of a theorem by V. Eberhard.
Math. Slovaca 27(1977), 383-407.

E. Jucovič

[1976] On face-vectors and vertex-vectors of cell-decompositions
of orientable 2-manifolds.
Math. Nachr. 73(1976), 285-295.

J. Kraeft

[1977] Über 3-realisierbare Folgen mit beliebigen Sechseckzahlen.
J. of Geometry 10(1977), 32-44.

149

K. Kuratowski

  [1930]   Sur le problème des courbes gauches en topologie.
              Fund. Math. 15(1930), 271-283.

S. Mac Lane

  [1937]   A combinatorial condition for planar graphs.
              Fund. Math. 28(1937), 22-32.

P. Mani

  [1971]   Automorphismen von polyedrischen Graphen.
              Math. Ann. 192(1971), 279-303.

P. Mani-Levitska, B. Guigas and W. E. Klee

  [1979]   Rectifiable n-periodic maps.
              Geometriae Dedicata 8(1979), 127-137.

R. Schmidt

  [1980]   (to appear)

S. K. Stein

  [1951]   Convex maps.
              Proc. Amer. Math. Soc. 2(1951), 464-466.

E. Steinitz

  [1961]   Polyeder und Raumeinteilungen.
              Enzykl. Math. Wiss. Vol. 3 (Geometrie), Part 3AB12,
              1-139 (1916).

E. Steinitz and H. Rademacher

  [1934]   Vorlesungen über die Theorie der Polyeder.
              Springer, Berlin 1934.

C. Thomassen

  [1977]   Straight line representations of infinite planar graphs.
              J. London Math. Soc. 16(1977), 411-423.

  [1980]   Planarity and duality of finite and infinite graphs.
              J. Combinat. Theory (B) 29(1980), 244-271.

  [1980*]   Planarity and duality of graphs.
              (to appear)

W. T. Tutte

  [1958]   Matroids and graphs.
              Trans. Amer. Math. Soc. 88(1958), 144-174.

  [1960]   Convex representations of graphs.
              Proc. London Math. Soc. 10(1960), 304-320.

  [1963]   How to draw a graph.
              Proc. London Math. Soc. 10(1963), 743-768.

K. Wagner

  [1936]   Bemerkungen zum Vierfarbenproblem.
              Jber. Deutsch. Math. Verein. 46(1936), 26-32.

  [1966]   Einbettung von überabzählbaren, simplizialen Komplexen
              in euklidische Räume.
              Arch. Math. 17(1966), 169-171.

  [1967]   Fastplättbare Graphen.
              J. Combinat. Theory 3(1967), 326-365.

H. Whitney

[1933]   Planar graphs.
           Fund. Math. 21(1933), 73-84.

Addresses of the authors

B. Grünbaum, Department of Mathematics, University of Washington,
     Seattle, Wa. 98195, U.S.A.

G. C. Shephard, University of East Anglia, Norwich NR4 7TJ.

SOME CONNECTIONS BETWEEN DESIGNS AND CODES

F.J. MacWilliams
Bell Laboratories,
Murray Hill,
New Jersey.

I: INTRODUCTION

This paper describes how to get designs from codes.

II: DEFINITIONS OF A DESIGN

Let x be a set of n points. A block is a subset of k points of x. A t-design is a collection of distinct blocks with the property that any t-subset of x is contained in exactly λ blocks. This will be called a t-(n,k,λ) design.

The parameters of a design needed for the first part of this talk are as follows:

Let $z = P_1 P_2 \ldots P_t$ be a t-subset of x. Let $\lambda^{(k)}_{i,t-i}$ be the number of blocks which contain i points of z, and do not contain the other $t - i$ points. $\lambda^{(k)}_{i,t-i}$ depends only on i, not on the choice of the i points. Thus the number of blocks containing exactly i points of z is $\binom{t}{i} \lambda^{(k)}_{i,t-i}$.

Definition of a code

Let $F = GF(2)$, F^n a vector space of dimension n over F. An (n,M) code c is a set of M vectors of F^n, including $\underline{0}$. If c is a subspace of F^n it is called a linear code, and written $[n,k]$, where $M = 2^k$. It is assumed that no coordinate place is always zero in all code words. The distance between two vectors is the number of places in which they differ: the weight of a vector is the number of ones it contains. Thus

$$\text{distance } (c_s, c_t) = \text{weight } (c_s + c_t).$$

$(a_0 a_1 \ldots a_n)$ is the distance distribution of c; that is

$$a_i = \frac{1}{M} \text{ (number of pairs } c_s, c_t \text{ such that weight } (c_s + c_t) = i)$$

If c is linear, $(a_0 a_1 \ldots a_n)$ is the weight distribution of c.
Let c_w = set of codewords of c of weight w.
$c^{(w)}$ = set of codewords of c of weight $\geq w$.
In the following there are given some sufficient conditions for c_w to be a t-design.

For any vector z of F^n, $A(z,k)$ is the number of $c \in C$ such that distance $(z,c) = k$, and $A^{(w)}(z,k)$ is the number of $c \in C^{(w)}$ such that distance $(z,c) = k$.

C is t-constant if $A(z,k)$ is the same for all z of weight t. It is assumed that $t \le d$.

Theorem 1. C_w is a t-design if and only if C is τ-constant, $0 \le \tau \le t$.

Proof (i) Assume C_w a t-design. Let z be a vector weight $\tau \le t$.

For $c \in C_w$ distance $(z,c) = k$ if and only if $|z \cap c| = \dfrac{w + \tau - k}{2}$. (Recall that distance $(z,c) = $ weight $(z+c) = $ weight $(z) + $ weight $(c) - 2$ weight $|z \cap c|$.)

Hence the number of $c \in C_w$ such that distance $(z,c) = k$ is

$$\begin{bmatrix} \tau \\ \dfrac{w + \tau - k}{2} \end{bmatrix} \lambda^{(w)}_{\frac{w+\tau-k}{2}, \frac{\tau+k-w}{2}} = s(z,k)$$

The number of $c \in C$ such that distance $(z,c) = k$ is

$$A(z,k) = \sum_{w=0}^{n} s(z,k).$$

which depends only on τ, k. Hence C is τ-constant, $0 \le \tau \le t$.

(ii) Assume C to be τ-constant. The proof is by induction. $A(z,k) = A^{(0)}(z,k)$ depends only on weight (z) and k for weight $z \le t$. Assume that $A^{(w)}(z,k)$ has this property. It will be shown that this implies that C_w is a t-design, and that $A^{(w+1)}(z)$ has the property.

It may be assumed that C_w is not empty; otherwise there is nothing to prove. Recall that $t \le d \le w$. Let z be a vector of weight t. The number of $c \in C_w$ which cover z is $A^{(w)}(z,w-t)$, which is constant by assumption. Hence (C_w) is a t-design.

Now let z be of weight $\tau \le t$. The number of $c \in C_w$ which cover exactly i ones of z is

$$\begin{bmatrix} \tau \\ i \end{bmatrix} \lambda^{(w)}_{i, t-i}$$

For such c, distance $(z,c) = w + \tau - 2i$. Thus

$$A^{(w+1)}(z, w + \tau - 2i) = A^{(w)}(z, w + \tau - 2i) - \begin{bmatrix} \tau \\ i \end{bmatrix} \lambda^{(w)}_{i, \tau-i}$$

$$A^{(w+1)}(z,k) = A^{(w)}(z,k) \quad k \ne w + \tau - 2i.$$

Thus $A^{(w+1)}(z,k)$ has required property and C is t-constant.

Q.E.D.

Lemma 2. If C is t-constant, $t \le d$, then $t \le d - \lceil \dfrac{d-2}{t} \rceil$

Proof. Assume $2t > d$, otherwise there is nothing to prove. Let α, β be vectors of weight t, with $|\alpha \cup \beta| = d - 1$, $|\alpha \cap \beta| = i$. Then $i = 2t - d - 1$. C_d is a t-design; hence there exists $\bar{\alpha}, \bar{\beta}$ in C_d such that $\bar{\alpha} > \alpha \quad \bar{\beta} > \beta \quad \alpha \ne \beta$.

$$|\bar{\alpha} \cap \bar{\beta}| \geq |\alpha \cap \beta| = i$$
distance $(\bar{\alpha},\bar{\beta}) \leq 2d - 2i = 4d - 4t + 2 \geq d.$

III

This section is to introduce some more notation. c is an (n,M,d) code, distance distribution $(a_0 a_1 \ldots a_n)$, where $a_0 = 1$ $a_1 = a_2 = a_{d-1} = 0$, $a_d \neq 0$.

s = number of nonzero a_i.
Let $P_k^{(n)}(x)$ be the Krawtchouk polynomial.

$$P_k(n)(x) = \sum_{j=0}^{k} (-1)^j \binom{x}{j} \binom{n-x}{k-j} \quad k = 0,1,\ldots,n.$$

Define $b_k = M^{-1} \sum_{i=0}^{n} a_i P_k(i).$

$b_0 = 1$ $b_1 = b_2 \ldots = b_{d'-1} = 0$ $b_{d'} \neq 0$
s' = number of nonzero b_i.
If c is linear, $(b_0 b_1 \ldots b_n)$ is the distance distribution of the dual code $c^\perp$. In any case, c is an orthogonal array of strength $d' - 1$.
Define

$$B(z,k) = M^{-1} \sum_{i=0}^{n} A(z,i) P_k(i) \tag{1}$$

Lemma 3. $b_k = 0$ implies that $B(z,k) = 0$ all z.
For proof see The Theory of Error-Correcting Codes, MacWilliams and Sloane.
Let $\alpha(x)$ be a polynomial of degree r, and expand $\alpha(x)$ in terms of Krawtchouk polynomials.

$$\alpha(x) = \sum_{k=0}^{r} \alpha_k P_k(x) \; ; \; 2^n \alpha_k = \sum_{i=0}^{r} \alpha(i) P_i(k). \tag{2}$$

Lemma 4. $M \sum_{k=0}^{n} \alpha(k) B(z,k) = 2^n \sum_{i=0}^{n} \alpha(i) A(z,i).$

Proof. Straightforward substitution.

IV.

Some sufficient conditions for c_w to be a t-design.
Theorem 5. Suppose $s \leq d'$. If $0 \leq t \leq d' - s$, then c_w is a t-design for all w.
Proof. Let z be a vector weight $t = d' - s$.
Let $c' = \{c-z: c \in C, c \supset z\}$ \hfill (1)
$\lambda^{(w)}(z)$ = number of $c \in c_w$ which contain z.

$$= A^{\prime}(0,w-t)$$

= number of $c^{\prime} \epsilon C^{\prime}$ of weight $w - t$.

$C^{\prime}$ is an orthogonal array of strength $d^{\prime} - t - 1$, which implies $b_1^{\prime} = b_2^{\prime} = \ldots b_{d^{\prime}-t-1}^{\prime} = 0$, or $B^{\prime}(0,1) = B^{\prime}(0,2) = \ldots = B^{\prime}(0,d^{\prime}-t-1) = 0$ and $B^{\prime}(0,0) = 1$. From (1)

$$M^{\prime}\delta_{0k} = \sum_{i=0}^{n-t} A^{\prime}(0,w-t)P_k^{(n-t)}(w-t),$$

$$= \sum_{i=0}^{n-t} \lambda^{(w)}(z)P_k^{(n-t)}(w-t), \quad 0 \le k \le d^{\prime} - t = s - 1.$$

Since C has s nonzero weights there are s unknown $\lambda^{(w)}(z)$. There are s linear equations, and the matrix of coefficients is invertible - this is a property of Krawtchouk polynomials. Hence there is a unique solution for $\lambda^{(w)}(z)$, independent of the choice of z of weight t. Thus C_w is a t-design.

Q.E.D.

Theorem 6. If C contains the all-one word j, then C_w is a t-design for $t \le d^{\prime} - s + t$.

Proof. by example.

Consider the Nordstrom-Robinson code. This is a (16, 256, 6) code with weights 6, 8, 10, 16 $d = d^{\prime} = 6$ $s = s^{\prime} = 4d^{\prime} - s + 1 = 3$. Form the equations of Theorem 5. These are

$$\lambda^{(6)} + \lambda^{(8)} + \lambda^{(10)} + \lambda^{(16)} = 32$$

$$7\lambda^{(6)} + 3\lambda^{(8)} - \lambda^{(10)} - 13\lambda^{(16)} = 0$$

$$18\lambda^{(6)} - 2\lambda^{(8)} - 6\lambda^{(10)} + 78\lambda^{(16)} = 0.$$

However $\lambda^{(16)} = 1$; hence may solve equations. $\lambda^{(6)} = 4$, $\lambda^{(8)} = 3$, $\lambda^{(10)} = 24$.

Q.E.D.

Cor 7. If $d > 2$, $d^{\prime} - s \le d - \lceil \frac{d-2}{4} \rceil$. This follows immediately from Lemma 2.

Theorem 8. Suppose $s^{\prime} \le d$. If $0 \le t \le d - s^{\prime}$ then C is t-constant.

Proof. Let z have weight $t = d - s^{\prime}$. It is required to show that $A(z,k)$ depends only on t,k. Let $\alpha(x)$ be a polynomial of degree max $(k,s^{\prime})$ such that $\alpha(i) = 0$ for $b_i \ne 0$. Expand $\alpha(x)$ as

$$\alpha(x) = P_k(x) + \sum_{j=0}^{s^{\prime}-1} \alpha_{kj}P_j(x).$$

Since $\alpha(i) = 0$ if $B(z,i) \ne 0$, $\alpha(k)B(z,k) = 0$ for $k \ne 0$. From (2)

$$M\alpha(0)B(z,0) = 2^n[A(z,k) + \alpha_{k0}A(z,0) + \ldots + \alpha_{s^{\prime}-1}A(z,s^{\prime}-1)].$$

If $c \epsilon C$ and distance $(z,c) \le s^{\prime} - 1$, then

$$weight\ c \le t + s^{\prime} - 1 = d - s^{\prime} + s^{\prime} - 1 = d^{\prime} - 1,$$

thus $c = 0$, distance $(z,c) =$ weight $(z) = t$

$$M\alpha(0) = 2^n[A(z,k) + \alpha_{k,t}]$$

This determines $A(z,k)$ as a function of t,k, and c is t-constant.

Q.E.D.

Theorem 9. If C is an even weight code, $2 \le s' \le d$ then C is t-constant for $r \le d' - s + 1$. No proof given.

Examples. The [7,3,4] simplex code, $d = 4$, $s' = 2$ hence the code vectors form a 2-design.

0	1	2	3	4	5	6
1	1	1	0	1	0	0
0	1	1	1	0	1	1
0	0	1	1	1	0	1
1	0	0	1	1	1	0
0	1	0	0	1	1	1
1	0	1	0	0	1	1
1	1	0	1	0	0	1

$2 - (7,4,2)$

The [24,12,8] extended Golay code. This has weights 8, 12, 16, 24. $d = d' = 9$, $s = s' = 4$. Thus either by theorem 6 or theorem 9, the codewords of each weight form a 5 design. In particular the codewords of weight 8 form a $5 - (24,8,1)$ design.

V: THE CONNECTION WITH GROUPS

Another way of getting designs from codes is as follows: The automorphism group of a code, $Aut(C)$, is the group of permutations of coordinate places which sends codewords into codewords. If this group is t-transitive or t-homogeneous (that is any t places can be mapped onto any other t-places) then of course the codewords of each weight form a t-design. Unfortunately very little is known about the automorphism groups of codes. It can be summarized in a few sentences. It is known that the [24,12,8] extended Golay code has automorphism group M_{24} - the largest Mathieu group - which is 5-fold transitive, so as already proved, the codewords of each weight form 5-designs.

The Reed-Muller codes are invariant under the general affine group, $GA(m)$. This group is doubly transitive, hence the codewords of each weight of a Reed-Muller code form a 2-design. It is also known that the automorphism of an extended quadratic residue code $[p+1, (p+1)/2]$ contains the group $PSL(2,p)$. This is always doubly transitive, hence the codewords of each weight form a 2-design. Further, if -1 is not a square in $GF(p)$ this group is 3-homogeneous, and the codewords of each weight form a 3-design.

COUNTING GRAPHS WITH A DUALITY PROPERTY

R.W. Robinson
Michigan State University
and
University of Newcastle

Abstract

The enumeration of graphs and other structures satisfying
a duality condition, such as self-complementarity, is surveyed.
A modification of Burnside's lemma due to de Bruijn is presented
in order to unify and simplify the treatment of such problems.
Some new equalities between classes of graphs and digraphs are
found which seem not to be explained by natural 1-1 correspondences.
Also some new natural 1-1 correspondences are derived using the
modified Burnside lemma. Asymptotic analyses of the exact
numbers are reviewed, and some recent results described.

1. Introduction

Methods for counting graphs and related structures which
satisfy a duality condition are well-established in the literature.
A duality condition is defined by invariance up to isomorphism
under some operation. Complementation is an operation which
has often been considered. Self-complementary structures
enumerated include graphs and digraphs [Re63], tournaments [Sr70],
n-plexes [Pa73a], m-ary relations [Wi74], multigraphs [Wi78],
eulerian graphs [Ro69], bipartite graphs [Qu79], sets [B59 and
B64], and boolean functions [Ni59, Eℓ60, Ha63, Ha64, and PaR-A].
Closely related are 2-colored or signed structures invariant
under color interchange or sign interchange. These take in
2-colored graphs [HP63 and Ha79], graphs in which points,
lines, or points and lines are signed [HPRS77], signed graphs
under weak isomorphism [So80], 2-colored polyhedra [R27 and

KnPR75], and necklaces [PS77 and Mi78]. The converse of a
digraph results when all orientations of arcs are reversed.
Self-converse digraphs have been counted with or without loops
[HP66b], without symmetric pairs of arcs [Sr70], and with
symmetric pairs of arcs as an enumeration parameter [Sr76].
Invariance under reversal of a spacial orientation is called
achirality. Achiral structures counted include necklaces
[PaR-Pa], plane trees [HR75], stereoisomerism classes of trees
[RoHB76], and colored polyhedra and boolean functions [PaR-Pa].
These operations can be combined in many cases; for instance
digraphs invariant under conversion followed by complement
are enumerated in [Pa73b].

In each of the examples mentioned there is a group G of
possible isomorphisms, and counting G-inequivalent structures
is accomplished using Burnside's lemma. In addition, the
operation ρ defining the duality property satisfies $\rho G = G \rho$
and $\rho^2 \in G$. It follows that the G-equivalence classes which
are not self-dual under ρ are paired up into $(G \cup \rho G)$-
equivalence classes, whereas self-dual G-equivalence classes
are also $(G \cup \rho G)$-equivalence classes. Thus the number of
self-dual classes is twice the number of $(G \cup \rho G)$-equivalence
classes, minus the number of G-equivalence classes. This
pattern was followed in justifying the results in all of the
examples mentioned of counting self-dual structures. A possible
exception is Redfield, who stated a general formula for the
number of 2-colorings invariant under color reversal and applied
it to 2-colorings of the vertices of the cube and the icosahedron
with respect to the rotation groups. He did not consider it
necessary to supply a proof.

In the next section it will be shown how to derive all
of these counting results for self-dual structures more
directly. The method relies on a modification of Burnside's

lemma which is essentially due to de Bruijn. In the following
four sections details are given for selections which are not
1-1, achiral necklaces and key rings, self-complementary sets,
and self-complementary digraphs. Several cases of apparently
accidental equality between the cardinalities of different classes
have been observed for self-complementary graphs, digraphs and
relations. These are discussed in Section 7, where some new
examples are provided. In the section after, the modified
Burnside lemma is used to provide some natural 1-1 correspondences
between graphs and digraphs of different sorts. In Section 9
asymptotic analyses of the exact numbers are summarized and some
recent results in the area are reported. In the final section
related results and unsolved problems are discussed.

2. Basic counting technique

First a basic counting lemma is presented, which is a
variation of Burnside's lemma. It is then shown how to apply
the lemma to count graphs with various duality properties.
It is also used to derive structural information on self-
complementary digraphs and tournaments.

Burnside's lemma applies to any finite group G represented
(faithfully or not) as permutations of a finite set X. It
gives the number N(G) of orbits of G on X as

$$(2.1) \qquad N(G) = \frac{1}{|G|} \sum_{g \in G} |\{x \in X \mid gx = x\}| \ .$$

This fact appears as Theorem VII in [Bull, p.191], but it was
known to Burnside [W76 and W-A] to have appeared earlier in work
of Frobenius. In fact, the idea has been traced back to Cauchy
[Ne79]. The required variation applies when an additional
permutation ρ on X is supplied such that $\rho G = G\rho$. Then
ρ is well-defined on the orbits of G. We denote by $N(G,\rho)$
the number of orbits left fixed by ρ, and obtain an expression

for $N(G,\rho)$ which reduces to (2.1) when ρ is the identity on X.

<u>Counting Lemma</u>. $N(G,\rho) = \dfrac{1}{|G|} \displaystyle\sum_{g\in G} |\{x \in X \mid g\rho x = x\}|$.

This lemma is implicit in [B63 and B67]. It is a special case of a more general version of Burnside's lemma which was stated in [B71] and proved (in even more generality) in [B79]. We outline a proof here for completeness and because the idea is easier to see in an uncluttered setting.

<u>Proof of Counting Lemma</u>. Suppose that x and y in X are G-equivalent, so that $x = gy$ for some $g \in G$. Then $\rho x = \rho gy$, and by hypothesis $\rho g = h\rho$ for some $h \in G$. Thus $\rho x = h\rho y$, i.e., ρx and ρy are G-equivalent. So ρ respects the G-equivalence classes of X.

Now consider the number of pairs (x,g) with $x \in X$ and $g \in G$ such that $g\rho x = x$. If the count is arranged over $g \in G$ initially then the sumation of the lemma is obtained. By arranging the count over $x \in X$ initially we can also show the number to be $|G| \cdot N(G,\rho)$ and thus establish the equality.

If x belongs to such a pair then $\rho x = g^{-1}x$ for some $g \in G$, so that ρ leaves the orbit of x fixed. In that case, let H denote the stabilizer of x, i.e., the subgroup of G consisting of all elements leaving x fixed. Then $hg\rho x = x$ just if $hx = x$, so x is a member of exactly $|H|$ of the pairs to be counted. Each of the other members of x's orbit contributes the same number of pairs, since the stabilizer subgroups are all conjugate to H. The size of x's orbit is $|G|/|H|$, so in all that orbit contributes $|G|$ pairs. $\quad\square$

3. Selections which are not 1-1

A selection of $m \geq 2$ elements from some finite store
R (repetitions allowed) can be thought of as a function from
$\{1,2,\cdots,m\}$ into R. Selections which are not 1-1 correspond
to equivalence classes under the alternating group A_m which
are left fixed by some odd permutation ρ. The symmetric
group S_m is just $A_m \cup \rho A_m$, so $\rho A_m = A_m \rho = S_m - A_m$. By
the Basic Counting Lemma, the number sought is the average
number of functions fixed by members of $S_m - A_m$.

A permutation with exactly σ_i cycles of length i in its
disjoint cycle decomposition for $1 \leq i \leq m$ is said to have
cycle type $(\sigma_1, \sigma_2, \cdots, \sigma_m)$. In order to be a permutation in
S_m the cycle type must satisfy

(3.1)
$$\sum_{i=1}^{m} i\sigma_i = m ,$$

and in order to be in A_m it must also have

(3.2)
$$\sum_{i=1}^{\lfloor m/2 \rfloor} \sigma_{2i} \equiv 0 \pmod{2} .$$

The number of permutations in S_m with cycle type $(\sigma_1, \sigma_2, \cdots, \sigma_m)$
is just

(3.3)
$$m! / \prod_{i=1}^{m} \sigma_i ! \, i^{\sigma_i} .$$

The number of functions left fixed by such a permutation is

(3.4)
$$|R|^{\sum \sigma_i} ,$$

since these are precisely the functions which respect the
cycle structure of the permutation. Taking the average, we
have

(3.5)
$$2 \sum{}' \; |R|^{\sum \sigma_i} / \prod_{i=1}^{m} \sigma_i ! \, i^{\sigma_i}$$

selections of m members of R which are not 1-1 where $\sum{}'$

denotes the sum over all cycle types satisfying (3.1) and (3.2).

In many applications the range set R is provided with
a weight function, say mapping R into $\{0,1,2,\cdots\}$. A
selection is then weighted according to the sum of the weights
of its elements with their multiplicities. The reasoning which
led to (3.4) then provides the ordinary generating function by
weight for selections which are not 1-1. Let $f(x)$ be the generating
function by weight for the range set. By analogy with (3.4)
the generating function for those functions left fixed by a
permutation of cycle type $(\sigma_1,\sigma_2,\cdots,\sigma_m)$ is just

(3.6)
$$\prod_{i=1}^{m} f(x^i)^{\sigma_i}.$$

Averaging over the $m!/2$ elements of $S_m - A_m$ gives

(3.7)
$$2 \sum{}' \prod_{i=1}^{m} f(x^i)^{\sigma_i}/\sigma_i! i^{\sigma_i}$$

for the generating function for selections which are not 1-1,
with $\sum'$ denoting the same sum as in (3.5).

The result (3.7) was obtained by Pólya [P37, pp.172-173].
His method was to count functions under S_m and A_m, following
the general outline given in the Introduction. Actually, Pólya
observed that there are direct derivations, using nothing about
groups, for formulas giving the numbers of all functions into
R and the numbers of those functions which are 1-1. By keeping
track of the domain size with an additional variable, one can
use Pólya's observations to obtain (3.7) in a very roundabout
way which does avoid group theoretic considerations.

4. Achiral necklaces and key rings

Consider a necklace of $m > 2$ beads to be allowed all
rotational symmetries but prevented from reflection by
embedding in the plane or some other device. A k-coloring is

an assigment of beads to colors from the store $\{c_1, c_2, \cdots, c_k\}$, with no restriction on the colors assigned to adjacent beads. These would be enumerated by applying Burnside's lemma to the cyclic group C_m acting on the k^m colorings which are available when the beads are all kept fixed. A necklace is <u>achiral</u> if a reflection of it is rotationally equivalent to the original. If ρ is some reflection of the necklace then the dihedral group D_m is just $C_m \cup \rho C_m$, so $\rho C_m = C_m \rho = D_m - C_m$. Thus by the Counting Lemma the number of achiral k-colored necklaces is the average number left fixed by members of $D_m - C_m$. These permutations can all be realized on the vertices of a regular plane m-gon by a reflection through a line of symmetry. Thus for odd m all leave a single bead fixed and transpose the rest in pairs. For even m, half of the reflections transpose all of the beads in pairs while half fix two opposite beads and transpose the rest in pairs. For a necklace to be fixed under such an operation only requires that transposed beads be assigned the same color. Thus we have k^n achiral colorings when $m = 2n - 1$ and $(k^{n+1} + k^n)/2$ when $m = 2n$.

Specific information on the number of colors of each sort used can be obtained by employing generating functions. For instance if $k = 2$ then we only have to allow a factor of $1 + x^2$ for each transposition and $1 + x$ for each fixed bead. This gives $(1 + x)(1 + x^2)^{n-1}$ when $m = 2n - 1$ and $(1 + x + x^2)(1 + x^2)^{n-1}$ when $m = 2n$ for the ordinary generating function of achiral 2-colored necklaces according to the number of beads assigned to some particular color.

In general, if the beads of a necklace are assigned to members of a weighted range set then achiral assignments can be enumerated by total weight in terms of the generating function $f(x)$ for the range set. Averaging fixed assignments over the elements of $D_m - C_m$ gives $f(x)f(x^2)^{n-1}$ when

$m = 2n - 1$, and $(f(x)^2 + f(x^2))f(x^2)^{n-1}/2$ when $m = 2n$,

for the generating function for achiral assignments. What is essential here is that reflections of the necklace have no effect on the elements of the range.

A different situation is presented by a key ring, on which each key can be affixed in one of two orientations. Again the rotational equivalence is given by C_m in case there are m keys on the ring. However the action of a reflection reverses the orientation of each key. Suppose that k patterns of key are available, and duplicates are allowed on a key ring. Suppose further that exactly ℓ of the patterns are invariant under orientation reversal. In order to have a key ring invariant under a reflection it is clearly necessary that any fixed key be of one of these ℓ patterns. On the other hand transposed keys may use any pattern and its reverse. Thus we have ℓk^{n-1} achiral key rings with $2n - 1$ keys and $(\ell^2+k)k^{n-1}/2$ with $2n$ keys.

An example of the key ring construction can be found in achiral plane trees. A rooted plane tree can be thought of as a key ring composed of planted plane trees by taking the branches at the root point ordered cyclically by the embedding in the plane. Let $f(x)$ denote the ordinary generating function for planted plane trees by number of lines, and let $f^*(x)$ denote the generating function for those which are invariant under a reflection of the plane. That is, $f^*(x)$ counts achiral planted plane trees. Then the generating function $r_m(x)$ for achiral rooted plane trees in which the root point has degree m is found by averaging over $D_m - C_m$ as for key rings. The result is

$$(4.1) \quad r_m(x) = \begin{cases} f^*(x)f(x^2)^{n-1} & \text{if } m = 2n - 1, \\ (f^*(x)^2 + f(x^2))f(x^2)^{n-1}/2 & \text{if } m = 2n. \end{cases}$$

Steric trees offer an illustration of similar reasoning
in three dimensions. In a steric tree each point is an endpoint
or else has degree 4. The points of degree 4 are called carbon
points. The neighbors of a carbon point are arranged around it
like the vertices of a regular tetrahedron around its center.
The rotations of a tetrahedron give the group A_4 on the
vertices, while if reflections are allowed S_4 is obtained.
The reverse of a steric tree is generated by reversing the
orientations of the neighbors of each carbon point. Achiral
carbon-rooted steric trees are then counted, according to the
Counting Lemma, by averaging the fixed trees over $S_4 - A_4$.
If $s(x)$ is the generating function for all planted steric
trees and $s*(x)$ for those that are achiral, this gives

(4.2) $$\frac{1}{2} s*(x)^2 s(x^2) + \frac{1}{2} s(x^4)$$

for achiral carbon-rooted steric trees.

Key rings first appeared as charm bracelets in [St74],
where they were used to give a derivation of the number of
triangulations of a regular n-gon inequivalent under the dihedral
group.

The original derivations of the number of achiral
necklaces [PaR-Pa],and of (4.1) for achiral rooted plane trees
[HR75] involved the usual comparison between counting under
C_m-equivalence and D_m-equivalence. In the case of achiral
trees it has been found that with the help of certain natural
correspondences with sequences the numbers can be derived without recourse to any variation on Burnside's lemma [Wo78].
Similarly, a direct argument relying on specific properties of
C_m and D_m was found for the number of 2-colored necklaces
which are both achiral and self-complementary [PaS77]. The
original derivation of (4.2) for achiral carbon-rooted steric

trees compared A_4-equivalence classes with S_4-equivalence classes [RoHB76]. It should be noted that the forms of (4.1) and (4.2) differ slightly from the original because of changes in notation.

Other sorts of achiral configurations which have been counted include vertex 2-colorings of various polyhedra in 3-space and of the n-cube in n-space [PaR-Pa]. The orientation preserving isometries are called rotations, while the orientation reversing isometries are referred to as reflections. To count colorings which are achiral with respect to rotations it is not necessary to deal with either the rotation group or the full group of isometries. By the   Counting   Lemma it suffices to average the number of fixed colorings over the coset of reflections.

5. Sets equivalent to their complements

If G is a finite group represented as permutations on a finite set X, then we say that subsets S,T $\subseteq$ X are G-equivalent if S = g(T) for some g $\in$ G. A subset S of X is self-complementary if S and X - S are G-equivalent. For later application of the idea, it is better to work with functions from X into {0,1} letting a function f correspond to the subset $f^{-1}(1)$. Then f corresponds to a self-complementary set just if f is G-equivalent to 1 - f, i.e., to the function resulting when 0 and 1 are transposed in the range.

Let ρ denote the result of interchaning 0 and 1 in the range, so that $\rho(f) = 1 - f$ for any $f : X \to \{0,1\}$. For $g \in G$ let $\tilde{g}$ be the induced action on functions from X into {0,1}, so that $\tilde{g}(f)$ is the composition $f(g^{-1})$ for each such function. Denoting by $\tilde{G}$ this representation of G, we note that ρ commutes with every element and so $\rho\tilde{G} = \tilde{G}\rho$.

Self-complementary subsets of G are being viewed as functions
f for which ρf is $\tilde{G}$-equivalent to f. Let the number of
$\tilde{G}$-equivalence classes of such functions be written $\bar{s}(G)$. By
the   Counting Lemma, $\bar{s}(G)$ is the average number of
functions left fixed by the members of $\rho\tilde{G}$.

To find the number of functions left fixed by $\rho\tilde{g}$,
consider any cycle $(x_1 x_2 \cdots x_i)$ induced by g on X.
For f to be left fixed requires $f(x_k) = (\rho\tilde{g}f)(x_k)$ for each
$1 \leq k \leq i$. But $(\tilde{g}f)(x_k) = f(g^{-1}(x_k)) = f(x_{k-1})$, so
$f(x_k) = 1 - f(x_{k-1})$ for $1 \leq k \leq i$ with subscripts modulo i.
This will only be possible when i is even, since values of
O and 1 must alternate around the cycle. If i is even,
then there are 2 such assignments which are possible. This
criterion applies independently to the disjoint cycles induced
by g on X. Thus if g induces $E(g)$ even length cycles
and $Q(g)$ odd length cycles, then the number of functions
fixed by $\rho\tilde{g}$ is $0^{Q(g)}2^{E(g)}$. The number is O if there are
any odd cycles. Averaging over $\rho\tilde{G}$ gives

(5.1) $$\bar{s}(G) = \frac{1}{|G|} \sum_{g \in G \& Q(g)=0} 2^{E(g)} .$$

By choosing the appropriate permutation group G, the
relation (5.1) can be applied to enumerate a variety of self-
complementary structures, 2-colorings invariant under color
interchange, and signed configurations invariant under sign
reversal. It was first proved in its full generality by
de Bruijn [B59 and B64]. His method was to use Pólya's
Hauptsatz to count $\tilde{G}$-equivalence classes, and his own
generalization of Polya's theorem to count $(\tilde{G} \cup \rho\tilde{G})$-equivalence
classes. He later gave a direct proof [B63 and B67] based
on his generalization of Burnside's lemma. Independent
discoveries of (5.1) in the context of specific applications
included Redfield [R27] for the rotation groups of the cube

and the icosahedron, and Ninomiya [Ni59] and Elspas [El60]
for the isometry group of the n-cube. Use was made of de Bruijn's
formulation of the general principal by Read for counting self-
complementary graphs and digraphs [Re63], and by Harrison for
counting self-complementary boolean functions under the general
linear and affine groups [Ha63 and Ha64]. Harary and Palmer
[HP66a] provided a new group theoretic setting for and proof of
de Bruijn's theorem, and thereby a new framework for applying (5.1).
From one or another of these points of view, a number of further
counting applications were made, including: 2-colored graphs
invariant under color reversal [HP63 and Ha79]; polygons with
vertices 2-colored and a subset of the diagonals 2-colored, invariant
under color reversal [La69]; self-complementary tournaments [Sr70];
self-complementary n-plexes [Pa73a]; self-complementary m-place
relations [Wi74]; 2-colored polyhedra isometrically invariant
under color reversal [KnPR75]; self-negational graphs in which
points, lines, or points and lines have been signed [HPRS77];
2-colored necklaces invariant by the dihedral group under color
reversal [PaS77]; q-colored necklaces invariant by the
cyclic group under reversal of $\lfloor q/2 \rfloor$ pairs of colors
[Mi78]; self-complementary bipartite graphs [Qu79];
complete graphs with signed lines which are self-negational under
weak isomorphism [So80]. Self-complementary eulerian graphs were
originally enumerated directly using (5.1), but were found to be
in a natural 1-1 correspondence with ordinary self-complementary
graphs on one fewer points [Ro69]. This sort of correspondence
is discussed in Section 8.

A curious consequence of (5.1) relates self-complementary
subsets to ordinary subsets. The generating function for
subsets by number of elements is $\sum_{n=0}^{k} s_n x^n$, where $k = |X|$
and s_n is the number of G-inequivalent subsets of X with
cordinality n. By a weighted version of Burnside's lemma,

or equivalently by Pólya's counting theorem, it can be seen that

$$\sum_{n=0}^{k} s_n x^n = \frac{1}{|G|} \sum_{g \in G} \prod_{i=1}^{k} (1 + x^i)^{\sigma(i,g)}$$

where $\sigma(i,g)$ is the number of cycles of length i induced on X by g. Setting $x = -1$ we obtain 0 for any product on the right involving at least one cycle of odd length. For products in which every cycle has even length one has $2^{E(g)}$, so that by (5.1) the result is $\bar{s}(G)$. On the left one obtains an alternating sum, so that

$$(5.2) \qquad \bar{s}(G) = s_0 - s_1 + s_2 \mp \cdots + (-1)^k s_k .$$

This relation was discovered by Frucht and Harary [FrH74], and is often useful as a cross check on numerical results.

6. Self-complementary digraphs and relations

To count the number $\bar{d}_p$ of self-complementary digraphs on p points, we can apply (5.1) to the group $S_p^{<2>}$ obtained by the action of S_p on the ordered pairs of distinct members of $\{1,2,\cdots,p\}$. The digraph corresponding to a subset of the pairs has i adjacent to j just if $\langle i,j \rangle$ is included in the set. All that is needed is to calculate the cycle structure of $\tilde{g} \in S_p^{<2>}$ induced by $g \in S_p$. In fact this depends only on the cycle structure of g. If γ and δ are disjoint cycles of length i and j respectively, then the ij arcs from γ to δ are permuted by g in (i,j) different cycles, each of length $[i,j]$. Here (i,j) denotes the greatest common divisor of i and j, and $[i,j]$ is the least common multiple. The same is true of the arcs from δ to γ. The $i(i-1)$ arcs joining points of γ to each other are permuted in $i-1$ cycles, each of length i. In order that a positive contribution to (5.1) be made by $\tilde{g}$, it is necessary that g induce no cycles of arcs of odd length. That amounts to g having no cycles of odd length greater than 1, and at

For an even number $p = 2n$ of points, the permutation g must consist soley of even length cycles. If g has σ_i cycles of length $2i$, $1 \leq i \leq n$, then the number $c(\sigma_1,\cdots,\sigma_n)$ of cycles of arcs can be calculated from the considerations above, since every arc must join points in two distinct cycles of g or else points in the same cycle of g. Summing over all possibilities gives

$$(6.1) \quad c(\sigma_1,\cdots\sigma_n) = \sum_{1 \leq i < j \leq n} 4\sigma_i\sigma_j(i,j) + \sum_{1 \leq i \leq n} 2i^2\sigma_i^2 + (2i-1-2i^2)\sigma_i .$$

As noted in Section 3, the number of permulations of $2n$ with this cycle structure is $(2n)! / \prod_{i=1}^{n} \sigma_i!(2i)^{\sigma_i}$. To adjust the powers of 2 it is convenient to let

$$c'(\sigma_1,\cdots,\sigma_n) = c(\sigma_1,\cdots,\sigma_n) - \sum_{i=1}^{n} \sigma_i .$$

Then by (5.1) we have

$$(6.2) \qquad \bar{d}_{2n} = \sum 2^{c'(\sigma_1,\cdots,\sigma_n)} / \prod_{i=1}^{n} \sigma_i! i^{\sigma_i} ,$$

where the sum is over $\sigma_i \geq 0$ satisfying

$$n = \sum_{i=1}^{n} i\sigma_i .$$

For an odd number $p = 2n+1$ of points, the only difference is the addition of a single fixed point to each permutation. This gives rist to two additional cycles of arcs for each even length cycle of points. If we let

$$c^+(\sigma_1,\cdots,\sigma_n) = c(\sigma_1,\cdots,\sigma_n) + \sum_{i=1}^{n} \sigma_i ,$$

then (5.1) gives

$$(6.3) \qquad \bar{d}_{2n+1} = \sum 2^{c^+(\sigma_1,\cdots,\sigma_n)} / \prod_{i=1}^{n} \sigma_i! i^{\sigma_i}$$

with the same summation convention as (6.2).

A binary relation differs from a digraph in allowing pairs of equal points, which we can visualize as loops when present. Complementation removes any loops present and supplies any loops that are absent, so that a self-complementary relation must have an even number $p = 2n$ of points. The number $\bar{r}_{2n}$ of self-complementary relations is found by modifying the calculation for $\bar{d}_{2n}$ with the addition of a cycle of loops for each point cycle. Thus in this case (5.1) implies

$$(6.4) \qquad \bar{r}_{2n} = \Sigma \, 2^{c(\sigma_1, \cdots, \sigma_n)} / \prod_{i=1}^{n} \sigma_i ! i^{\sigma_i} \, ,$$

with the same summation as (6.2) and (6.3).

Expressions equivalent to (6.2) and (6.3) were first obtained by Read [Re63], and for (6.4) by Wille [Wi74]. The notation $S_p^{\langle 2 \rangle}$ for the ordered pair group is due to Harary; see Chapter 5 of the book [HP73] for a thorough discussion of its cycle structure and applications to counting digraphs. Most problems on counting self-complementary graphs, relations and multigraphs follow a pattern similar to the calculation of $\bar{d}_p$ and $\bar{r}_p$. The same applies to self-converse digraphs. However for self-converse oriented graphs, self-converse tournaments, and digraphs with complement isomorphic to converse (5.1) cannot be applied directly. This is because of the necessity of distinguishing for these problems between line cycles in which the endpoints of the lines are reversed, and line cycles in which the endpoints are not reversed. Even though the derivations in such cases must rely directly on Burnside's lemma or on its variant the Counting Lemma, the calculations are still quite similar to those for $\bar{d}_p$ and $\bar{r}_p$.

7. __Equinumerous classes of digraphs__

Read [Re63] proved the remarkable fact that the number $\bar{g}_{4n}$ of self-complementary graphs on $4n$ points is equal to

the number $\bar{d}_{2n}$ of self-complementary digraphs on $2n$ points.
Wille [Wi78] showed the parallel equality $\bar{g}_{4n+1} = \bar{r}_{2n}$. There
are no known natural 1-1 correspondences to explain these
equalities. We present a related equality which seems equally
inexplicable on structural grounds.

By a **bilayered digraph** we mean a superposition of two digraphs,
where the distinction is maintained between the arcs of the
upper digraph and the arcs of the lower digraph. In Figure 1
the ten inequivalent bilayered digraphs on 2 points are pictured,
the distinction between the layers indicated by dotted and
solid arcs.

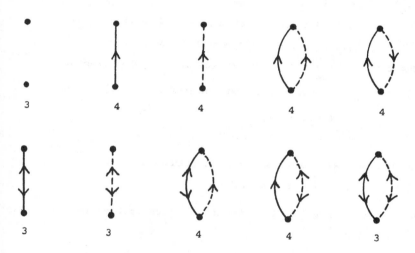

Figure 1 Bilayered digraphs on 2 points

We will show that the number b_n of nonisomorphic bilayered
digraphs on n points is equal to the number $\bar{d}_{2n}$ of self-
complementary digraphs on $2n$ points.

Theorem 1. $b_n = \bar{d}_{2n}$ for all $n \geq 1$.

Proof. A bilayered digraph can be thought of as a mapping from $S_n^{\langle 2 \rangle}$ into $\{0,1,2,3\}$, where the assignment of $\langle i,j \rangle$ to 0 denotes no arc from i to j, to 1 denotes a solid arc only from i to j, to 2 denotes a dotted arc only from i to j, and to 3 denotes both the solid and dotted arcs from i to j. Thus by Burnside's lemma the number of non-isomorphic bilayered digraphs is the average number of mappings left fixed by the elements of $S_n^{\langle 2 \rangle}$. If $g \in S_n$ has the cycle structure $(\sigma_1, \cdots \sigma_n)$ and $\tilde{g}$ induces $\beta(\sigma_1, \cdots \sigma_n)$ different cycles of arcs in $S_n^{\langle 2 \rangle}$, then $\tilde{g}$ leaves exactly $4^{\beta(\sigma_1, \cdots, \sigma_n)}$ mappings fixed. What is required is that such mappings be well-defined on the cycles of $\tilde{g}$, so there are 4 possible images for each cycle. There are $n! / \prod_{i=1}^{n} \sigma_i! i^{\sigma_i}$ permutations of cycle structure $(\sigma_1, \cdots \sigma_n)$ in S_n, so averaging over the members of $S_n^{\langle 2 \rangle}$ gives

(7.1)
$$b_n = \Sigma \, 4^{\beta(\sigma_1, \cdots, \sigma_n)} / \prod_{i=1}^{n} \sigma_i! i^{\sigma_i} \, ,$$

taking the sum over all possible cycle structures for S_n. That is, $\sigma_i \geq 0$ for $1 \leq i \leq n$ and $n = \Sigma \, i\sigma_i$.

From the discussion of $S_n^{\langle 2 \rangle}$ in the previous section it is easily seen that

(7.2) $\beta(\sigma_1, \cdots, \sigma_n) = \sum_{1 \leq i < j \leq n} 2\sigma_i \sigma_j (i,j) + \sum_{1 \leq i \leq n} i^2 \sigma_i^2 + (i-1-i^2)\sigma_i$.

Now from (6.1) and the definition of $c'(\sigma_1, \cdots, \sigma_n)$ we have

$$c'(\sigma_1, \cdots, \sigma_n) = \sum_{1 \leq i < j \leq n} 4\sigma_i \sigma_j (i,j) + \sum_{1 \leq i \leq n} 2i^2 \sigma_i^2 + (2i-2-2i^2)\sigma_i$$

$$= 2\beta(\sigma_1, \cdots, \sigma_n) \, .$$

Thus the $(\sigma_1, \cdots, \sigma_n)$ term in (7.1) equals the corresponding term in (6.2), so the sums b_n and $\bar{d}_{2n}$ are equal. $\square$

In Figure 2 the ten self-complementary digraphs on 4 points
are pictured. Together, Figures 1 and 2 illustrate the equality
just shown for n = 2.

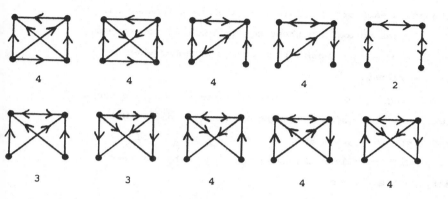

4   4   4   4   2

3   3   4   4   4

Figure 2   Self-complementary digraphs on 4 points

Since the proof of Theorem 1 establishes equality between
the contributions of corresponding cycle structures to b_n and
$\overline{d}_{2n}$, it can be lifted to accomodate signing and coloring the
points. For instance if each point is assigned either + or
- and one of k colors $\{1,2,\cdots,k\}$, then an additional
factor of 2k is required for each point cycle in computing
the number $b_n^{(2k)}$ of bilayered digraphs which are so marked.
For marked digraphs it is natural to consider those which are
invariant under simultaneous complementation of arcs and
reversal of signs. For an even number 2n of points this also
requires an additional factor of 2k for each point cycle in
computing $\overline{d}_{2n}^{(2k)}$ since the point cycle lengths must all be even
already to allow arc complementation. Thus we have the following
corollary.

  <u>Corollary 1</u>. $b_n^{(2k)} = \overline{d}_n^{(2k)}$ for all $n, k \geq 1$.

For $n = 2$ and $k = 1$ the numbers of signed versions
of the digraphs involved are shown in Figures 1 and 2. In
each case the total is 36, illustrating the equality given
in the corollary. However the individual numbers do not
correspond to each other. This indicates that finding natural
correspondences to explain these equalities will not be easy,
as one for $b_n = \bar{d}_{2n}$ cannot be directly extended when the
points are signed.

Replacing $+$ with a loop and $-$ with no loop provides an
obvious 1-1 correspondence showing that $\bar{r}_{2n} = \bar{d}_{2n}^{(2)}$. With
the corollary this gives an extension $\bar{r}_{2n} = b_n^{(2)}$ of Wille's
equality which parallels the extension $\bar{d}_{2n} = b_n$ of Read's
equality in the theorem.

There are several other identities in the literature
which do not seem to be explicable by natural 1-1 correspondences.
Schwenk [HPRS77, (9)] has found one showing that the number of
point-signed oriented graphs of order $2n$ with $(k-1)$-colored
arcs which are invariant under sign negation is the same
as the number of $(k^2 - 1)$-line-colored graphs of order n.
A closely related identity [HPRS77, (14)] shows that the
number of k-line-colored point-signed graphs of order $2n$
invariant under sign negation is the same as the number of k-arc
colored relations of order n. The number of switching classes
of order n is the same as the number of even graphs of order
n [MaS75]. For odd n there is a natural correspondence
between the two, but none has been discovered for even n.
In [Pa73a] it is noted that $\bar{d}_8 = 703,760$ is also the number
of self-complementary 2-plexes of order 8, an apparently
isolated phenomenon. Further families of equalities involving
digraphs and tournaments are explored in [Ro-Pb].

8. Some natural correspondences

We start by showing that any self-complementary digraph δ of odd order $2n+1$ can be rooted so as to remain self-complementary in exactly one way up to isomorphism. Let $\Gamma(\delta)$ denote the automorphism group of δ. For $g \in S_{2n+1}$ let $\bar{g}$ denote the result of g followed by complementation, and let $\Gamma'(\delta)$ be the set of all g such that $\bar{g}(\delta) = \delta$. From the discussion leading to equation (6.3) for $\bar{d}_{2n+1}$ it is clear that any such g must leave exactly one point of δ fixed. Thus the average number of fixed points over $\Gamma'(\delta)$ is 1. If we pick some $g \in \Gamma'(\delta)$ then $\Gamma'(\delta) = g\Gamma(\delta) = \Gamma(\delta)g$, so the Counting Lemma applies to show that there is exactly one similarity class F of points of δ which is left fixed by $\Gamma'(\delta)$. Since there are points which are individually left fixed by members of $\Gamma'(\delta)$, F consists precisely of those points. If δ_x denotes the result of rooting δ at the point x, then δ_x is self-complementary just if $x \in F$. Thus we have shown the following.

Theorem 2. Any self-complementary digraph of odd order has a unique point-rooted version which is self-complementary.

Corollary 2. $\bar{d}_{2n+1} = \bar{d}_{2n}^{(4)}$ for all $n \geq 1$.

Proof. The theorem allows the corollary to be deduced from the following natural correspondence. From a digraph δ of order $2n$ with signed and 2-colored points, form a digraph $\hat{\delta}$ of order $2n+1$ by adding a new point v. Any point u of δ is joined to v by an arc just if the sign of u is $+$. Conversely, v is joined to u just if the color of u is 1 and its sign is $+$, or else the color is 2 and its sign is $-$. The points of $\hat{\delta}$ are to have neither signs nor colors attached. Sign reversal on δ corresponds to complementation of the arcs to and from v, so δ is invariant under complementation

and sign reversal just if $\hat{\delta}$ is self-complementary by way of
a permutation which fixes v. The correspondence is obviously
1-1 if $\hat{\delta}$ is considered to be rooted at the new point, and by
Theorem 2 the rooting is unique for self-complementary digraphs
of odd order. □

There is an obvious correspondence between the digraphs
counted by $\overline{d}_{2n}^{(4)}$ and the self-complementary 2-colored relations
on 2n points. It is obtained from the correspondence for
$\overline{d}_{2n}^{(2)} = \overline{r}_{2n}$ given in the previous section by 2-coloring the
points on both sides. Thus it is seen that $\overline{d}_{2n+1}$ is also the
number of self-complementary 2-colored relations of order 2n.

The theorem applies to the special case of self-complementary
tournaments. The reasoning of the corollary in this case gives
the following by a natural 1-1 correspondence.

Corollary 3. The number of self-complementary tournaments
on 2n + 1 points is exactly the number of point-signed
tournaments on 2n points which are invariant under simultaneous
complementation of arcs and sign reversal.

It can be shown in a similar fashion that $\overline{g}_{4n+1}$ is the
number of self-complementary symmetric relations of order 4n.
The fact that every self-complementary graph of odd order has
a unique self-complementary rooting in the key, and that can
be established in the same way as Theorem 2. However it can
also be established in a much more elementary way, as in
[Ro69] where it was shown that $\overline{g}_{4n}$ is the number of self-
complementary even graphs of order 4n + 1. For a map of a graph
of odd order to its complement consists of one fixed point
along with cycles of length divisible by 4. The square is an
automorphism with one fixed point and the rest even cycles,
so one can identify the similarity class of fixed points of

omplementations by the fact that it is the only class of odd
ardinality. No such simple criterion holds for digraphs,
s illustrated in Figure 3. The self-complementary tournament
f the figure has score sequence (1,2,2,2,3) and identity
utomorphism group.

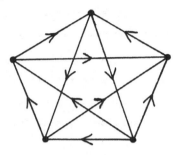

<u>Figure 3</u>   An identity self-complementary tournament

The structural fact necessary to establish Corollary 3
s implicit in [Ep79], where an appeal to the conjugacy of
-Sylow subgroups replaces our use of the Counting Lemma.
owever it appears that this method is not applicable to
elf-complementary digraphs.

. <u>Asymptotic analysis</u>

In unlabeled graph and digraph counting, it is commonly
he case that asymptotically almost all have identity
utomorphism group. Thus the number is asymptotic to the
ontribution of the single term given by the identity permutation
a applying Burnside's lemma. There are a number of self-dual
ounting problems for which it is also true that a single term
s dominant asymptotically, including: self-complementary

graphs and digraphs [Pa70]; self-complementary 2-plexes
[Pa73a]; digraphs with converse and complement isomorphic
[Pa73b]; point-signed graphs invariant under sign reversal,
point- and line-signed graphs invariant under reversal of
point signs only, and point-and line-signed graphs invariant
under reversal of all signs [HPRS77]; self-complementary m-
graphs [Wi78]; self-complementary tournaments [Sr78].

However there are a number of cases in which this simple
asymptotic behavior is not observed. Compared with the nearly
uniform behavior of the corresponding ordinary unlabelled
counting problems, the vagaries exhibited by self-dual counting
problems are striking.

Wille [Wi78] showed that the number $\overline{g}_{2k}(p)$ of self-
complementary 2k-graphs is asymptotic to the sum of the
contributions from permutations consisting only of fixed points
and transpositions of points. In a 2k-graph, the complement of
a bundle of m parallel lines is a bundle of 2k-m parallel
lines. Straightforward analysis of Wille's sum shows that the
maximum occurs in the vicinity of 2M(p) fixed points, where

$$(9.1) \quad M(p) = \begin{cases} \frac{1}{2}\log p - \log\log p + \frac{1}{2} & \text{when } p \text{ is even,} \\ \\ \frac{1}{2}\log p - \log\log p & \text{when } p \text{ is odd,} \end{cases}$$

and all logarithms are to base $2k+1$. Expanding around the
maximum and using Stirling's formula to estimate the factorials,
the dependence on the number of fixed points is resolved into
a factor of $\sum\limits_{-\infty}^{\infty} (2k+1)^{-(m(p)-y)^2}$. Asymptotically this can be
replaced by $\sqrt{\pi \log e}\ \Phi(M(p))$, where Φ is the function of period
1 and average 1 given by

$$(9.2) \qquad \Phi(z) = \frac{1}{\sqrt{\pi \log e}} \sum\limits_{-\infty}^{\infty} (2k+1)^{-(z-i)^2} .$$

In this way one calculates

$$(9.3) \qquad \bar{g}_{2k}(p) \sim (2k+1)^{p^2/4} \left(\frac{e}{(2k+1)p}\right)^{p/2} p^{(\log p)/4 - \log\log p + \log e}$$
$$\cdot (\log p)^{\log\log p - 3/2} (2k+1)^{1/4} \left(\frac{\log e}{2\pi}\right)^{1/2} \Phi(M(p)) .$$

Here all logarithms are base $2k+1$, $k \geq 1$ is fixed and $p \to \infty$.

The expression of (9.3) for $k = 1$ appeared in [HPRS77, (16)] as the asymptotic behavior of the number s_p^* of line-signed graphs of order p invariant under sign reversal. Since the complement of a single line is itself in a 2-graph (multigraph of strength 2), an exact correspondence is provided if a single line is replaced by no line, no line by a negative line and two parallel lines by a positive line. Thus $s_p^* = \bar{g}_2(p)$ for all $p \geq 1$. A very similar expression was found [HPRS77, (18)] for the asymptotic number of graphs of order p with both lines and points signed, invariant under reversal of signs on the lines. Another example of similar asymptotic behavior is the number of self-converse oriented graphs [Ro78].

An example of somewhat different behavior is provided by the number d_p' of self-converse digraphs. Again the asymptotic growth is dominated by contributions from cycle structures with only fixed points and transpositions. However, in this case the dependence on the number of fixed points exhibits a much broader peak and is asymptotic to a fixed integral. The net result is

$$(9.4) \qquad d_p' \sim \frac{2^{(p^2-p)/2}}{p!} \left(\frac{2p}{e}\right)^{p/2} \frac{e^{\sqrt{p/2}}}{e^{1/8\sqrt{2}}} .$$

This corrects an erroneous impression [Sr79] that d_p' can be accurately estimated from the contributions of just two cycle types. The details and numerical data will appear elsewhere [Ro-Pa]. Asymptotic expressions similar to (9.4) have been

found for self-complementary bipartite graphs [Qu-P] and
self-complementary boolean functions [PaR-Pb].

10.   Related results

One direction for extension of the counting results is to
require connectivity. It is well known that all self-complementary
graphs are connected, and hence all self-complementary digraphs
are weakly connected, as are digraphs with converse isomorphic
to complement. Separable self-complementary graphs are subject
to severe structural restrictions which allow them to be
enumerated also [AH-A]. Weakly connected self-converse digraphs
and oriented graphs cannot be enumerated in the usual way
from the unrestricted ones, contrary to [SrP72]. This is because
a minimal self-converse digraph can be formed from a distinct
pair of converse digraphs. However essentially the same difficulty
was overcome in [HP63] and [Ha79] in order to count connected
properly 2-colored graphs invariant under color interchange.
The same technique can be applied to weakly connected digraphs
and oriented graphs which are self-converse.

A direction for generalization is to require the number
of structures with a given automorphism group, or a variation
thereof. A self-duality requirement is just one example of
a restriction on the automorphism groups to be considered.
If the groups are sufficiently restricted complete information
can sometimes be obtained. An example of this is the determination
by Lucas of the nonattacking arrangements of n rooks on an $n \times n$
chessboard having each of the possible symmetry groups [Lu91].
Similar results have been obtained for arrangements of bishops [Ro76].
Other examples of complete information occur when the groups
are constrained to be cyclic [CuR76] or to be products of
symmetric groups [Os73]. Point symmetric graphs, digraphs
and tournaments on a prime number of points have also been

enumerated, in [Tu67], [Al70], [Ch71] and [ChW73]. This was
possible because the groups involved are well understood. There
have also been very general solutions, valid for all possible
groups, e.g. [Sh68], [St71], [Wh75] and [Kl76]. However these all
require the table of marks for S_p in order to deal with the
automorphism groups of degree p and so become unworkable for
very modest values of p.

A much less ambitious objective is to count graphs or other
structures subject to two separate duality conditions. However
this appears very difficult in general. A typical long-standing
unsolved problem is to count digraphs which are both self-converse
and self-complementary. In fact the only solved problem of this
type is the number of self-complementary achiral necklaces
[PaS77], which relies heavily on the fact that the groups are
all dihedral or cyclic.

A related problem is to count labeled structures which
satisfy some duality condition. In general this seems to require
a complete knowledge of the automorphism groups. Thus labeled
graphs on a prime number of points which are point and line
symmetric can be counted [Ch74] but labeled self-complementary
graphs or digraphs have so far resisted enumeration.

In a different direction, self-complementary graphs
and digraphs have been enumerated by degree sequence [PaS69].
The method is very cumbersome, but has proved of some use in
comparing results with a generation procedure [KrR79]. A
similarly unwieldy method has been proposed for enumerating
self-complementary digraphs and oriented graphs in which the
indegree and outdegree are equal at every point [SrP72].

Enumeration methods aim at being far more efficient than
the generation of all the structures under study. However
generation methods are much more widely applicable than

enumeration, and have benefited from recent theoretical advances.
Evidence of this can be found in the cataloguing of the self-
complementary graphs on 12 points, independently in [Fa78]
and [KrR79].

REFERENCES

[AH-A]   J. Akiyama and F. Harary, A graph and its complement
         with specified properties IV: counting self-
         complementary blocks, to appear.

[Aℓ70]   B. Alspach, On point-symmetric tournaments, Canad.
         Math. Bull. 13 (1970) 317-323.

[B59]   N.G. deBruijn, Generalization of Pólya's fundamental
         theorem in enumerative combinatorial analysis,
         Nederl. Akad. Wetensch. Proc. Ser. A 62 =
         Indag. Math. 21 (1959) 59-69.

[B63]   N.G. deBruijn, Enumerative combinatorial problems
         concerning structures, Nieuw Arch. Wisk. (3)
         11 (1963) 142-161.

[B64]   N.G. deBruijn, Pólya's theory of counting, Applied
         Combinatorial Mathematics (E.F. Beckenbach, ed.)
         Wiley, New York (1964) 144-184.

[B67]   N.G. deBruijn, Colour patterns which are invariant under
         a given permutation of the colours, J. Combin.
         Theory 2 (1967) 418-421.

[B71]   N.G. deBruijn, Recent developments in enumeration theory,
         Actes Congrès intern. Math. 1970 vol. 3,
         Gauthier-Villars, Paris (1971) 193-199.

[B79]   N.G. deBruijn, A note on the Cauchy-Frobenius lemma,
         Nederl. Akad. Wetensch. Proc. Ser. A 82
         = Indag. Math. 41 (1979) 225-228.

[Bu11]   W. Burnside, The Theory of Groups of Finite Order,
         2nd ed., Cambridge (1911); reprinted Dover,
         New York (1955).

[Ch71]   C.Y. Chao, On the classification of symmetric graphs
         with a prime number of vertices, Trans. Amer.
         Math. Soc. 158 (1971) 247-256.

[Ch74]   C.Y. Chao, A note on labeled symmetric graphs, Discrete
         Math. 8 (1974) 295-297.

[ChW73] C.Y. Chao and J.G. Wells, A class of vertex-transitive
         digraphs, J. Combin. Theory Ser. B 14 (1973)
         246-255.

[CuR76]   L.J. Cummings and R.W. Robinson, Linear symmetry classes, *Canad. J. Math.* 28 (1976) 1311-1319.

[El60]   B. Elspas, Self-complementary symmetry types of boolean functions, *IRE Trans.* EC-9 (1960) 264-266.

[Ep79]   W.J.R. Eplett, Self-converse tournaments, *Canad. Math. Bull.* 22 (1979) 23-27.

[Fa78]   I.A. Faradžev, The obtaining of a complete list of self-complementary graphs up to 12 vertices, *Algorithmic Studies in Combinatorics* (I.A. Faradžev, ed.) Izdat. "Nauka", Moscow (1978) 69-75, 186.

[FrH74]   R. Frucht and F. Harary, Self-complementary generalized orbits of a permutation group, *Canad. Math. Bull.* 17 (1974) 203-208.

[Ha79]   P. Hanlon, The enumeration of bipartite graphs, *Discrete Math.* 28 (1979) 49-57.

[HP66a]   F. Harary and E.M. Palmer, The power group enumeration theorem, *J. Combin. Theory* 1 (1966) 157-173.

[HP66b]   F. Harary and E.M. Palmer, Enumeration of self-converse digraphs, *Mathematika* 13 (1966) 151-157.

[HP73]   F. Harary and E.M. Palmer, *Graphical Enumeration*, Academic, New York (1973).

[HPRS77] F. Harary, E.M. Palmer, R.W. Robinson and A.J. Schwenk, Enumeration of graphs with signed points and lines, *J. Graph Theory* 1 (1977) 295-308.

[HP63]   F. Harary and G. Prins, Enumeration of bicolourable graphs, *Canad. J. Math.* 15 (1963) 237-248.

[HR75]   F. Harary and R.W. Robinson, The number of achiral trees, *J. Reine Angew. Math.* 278/279 (1975) 322-335.

[Ha63]   M.A. Harrison, The number of equivalence classes of boolean functions under groups containing negation, *IEEE Trans. Electronic Computers* EC-12 (1963) 559-561.

[Ha64]   M.A. Harrison, On the classification of boolean functions by the general linear and affine groups, *J. Soc. Indust. Appl. Math.* 12 (1964) 285-299.

[Kl76]   M.J. Klass, A generalization of Burnside's combinatorial lemma, *J. Combin. Theory Ser. A* 20 (1976) 273-278.

[KnPR75] O. Knop, E.M. Palmer and R.W. Robinson, Arrangements of charges having zero electric-field gradient, *Acta Cryst. Sect. A* 31 (1975) 19-31.

[KrR79]   M. Kropar and R.C. Read, On the construction of the self-complementary graphs on 12 nodes, *J. Graph Theory* 3 (1979) 111-125.

[La69]   C.P. Lawes, *Some applications of group theory to enumerative combinatorial analysis*, dissertation, Dartmouth Coll., Hanover, N.H. (1969).

[Lu91]   E. Lucas, Théorie des Nombres, Gauthier-Villars,
          Paris (1891); reprinted, Blanchard, Paris (1961).

[MaS75]   C.L. Mallows and N.J.A. Sloane, Two-graphs, switching
          classes and Euler graphs are equal in number,
          SIAM J. Appl. Math. 28 (1975) 876-880.

[Mi78]   R.L. Miller, Necklaces, symmetries and self-reciprocal
          polynomials, Discrete Math. 22 (1978) 25-33.

[Ne79]   P.M. Neumann, A lemma that is not Burnside's,
          Math. Scientist 4 (1979) 133-141.

[Ni59]   I. Ninomiya, On the number of genera of boolean
          functions of n variables, Mem. Fac. Engineering
          Nagoya Univ. 11 (1959) 54-58.

[Os73]   L.J. Osterweil, A theorem for enumerating certain
          types of collections, Canad. J. Math. 25 (1973)
          74-82.

[Pa70]   E.M. Palmer, Asymptotic formulas for the number of self-
          complementary graphs and digraphs, Mathematika 17
          (1970) 85-90.

[Pa73a]   E.M. Palmer, On the number of n-plexes, Discrete Math.
          6 (1973) 377-390.

[Pa73b]   E.M. Palmer, Graphical enumeration methods, New
          Directions in the Theory of Graphs (F. Harary, ed.)
          Academic, New York (1973) 187-206.

[PaR-A]   E.M. Palmer and R.W. Robinson, Enumeration of self-dual
          configurations, to appear.

[PaR-Pa] E.M. Palmer and R.W. Robinson, Enumeration of achiral
          configurations, in preparation.

[PaR-Pb] E.M. Palmer and R.W. Robinson, Asymptotic numbers of
          boolean functions, in preparation.

[PaS77]   E.M. Palmer and A.J. Schwenk, The number of self-
          complementary achiral necklaces, J. Graph Theory
          1 (1977) 309-316.

[PaS69]   K.R. Parthasarathy and M.R. Sridharan, Enumeration of
          self-complementary graphs and digraphs, J. Math.
          Phys. Sci. 3 (1969) 410-414.

[P37]   G. Pólya, Kombinatorische Anzahlbestimmungen für Gruppen,
          Graphen und chemische Verbindungen, Acta Math.
          68 (1937) 145-254.

[Qu79]   S.J. Quinn, Factorisation of complete bipartite graphs
          into two isomorphic subgroups, Combinatorial
          Mathematics VI (A. Horadam and W.D. Wallis, eds.)
          Lecture Notes in Math. 748, Springer, Berlin (1979)
          98-111.

[Qu-P]   S.J. Quinn, Asymptotic number of self-complementary
          bipartite graphs, in preparation.

[Re63] R.C. Read, On the number of self-complementary graphs and digraphs, *J. London Math. Soc.* 38 (1963) 99-104.

[R27] J.H. Redfield, The theory of group-reduced distributions, *Amer. J. Math.* 49 (1927) 433-455.

[Ro69] R.W. Robinson, Enumeration of euler graphs, *Proof Techniques in Graph Theory* (F. Harary, ed.) Academic, New York (1969) 147-153.

[Ro76] R.W. Robinson, Counting arrangements of bishops, *Combinatorial Mathematics IV* (L.R.A. Casse and W.D. Wallis, eds.) Lecture Notes in Math. 560, Springer, Berlin (1976) 198-214.

[Ro78] R.W. Robinson, Asymptotic number of self-converse oriented graphs, *Combinatorial Mathematics* (D.A. Holton and J. Seberry, eds.) Lecture Notes in Math. 686, Springer, Berlin (1978) 255-266.

[Ro-Pa] R.W. Robinson, Asymptotic number of self-converse digraphs, in preparation.

[Ro-Pb] R.W. Robinson, Equinumerous classes of graphs and digraphs, in preparation.

[RoHB76] R.W. Robinson, F. Harary and A.T. Balaban, The number of chiral and achiral alkanes and monosubstituted alkanes, *Tetrahedron* 32 (1976) 355-361.

[Sh68] J. Sheehan, The number of graphs with a given automorphism group, *Canad. J. Math.* 20 (1968) 1068-1076.

[So80] T. Sozański, Enumeration of weak isomorphism classes of signed graphs, *J. Graph Theory* 4 (1980) 127-144.

[Sr70] M.R. Sridharan, Self-complementary and self-converse oriented graphs, *Nederl. Akad. Wetensch. Proc.* Ser. A 73 = *Indag. Math.* 32 (1970) 441-447.

[Sr76] M.R. Sridharan, Mixed self-complementary and self-converse digraphs, *Discrete Math.* 14 (1976) 373-376.

[Sr78] M.R. Sridharan, Note on an asymptotic formula for a class of digraphs, *Canad. Math. Bull.* 21 (1978) 377-381.

[Sr79] M.R. Sridharan, Asymptotic formula for the number of self-converse digraphs, *Proc. Symp. Graph Theory (Indian Statist. Inst., Calcutta, 1976)* (A.R. Rao, ed.) Macmillan of India, Dehli (1979) 299-304.

[SrP72] M.R. Sridharan and K.R. Parthasarathy, Isographs and oriented isographs, *J. Combin. Theory Ser. B* 13 (1972) 99-111.

[St71] P.K. Stockmeyer, *The enumeration of graphs with prescribed automorphism group*, dissertation, Univ. Michigan, Ann Arbor (1971).

[St74] P.K. Stockmeyer, The charm bracelet problem and its applications, Graphs and Combinatorics (R.A. Bari and F. Harary, eds.) Lecture Notes in Math. 406, Springer, Berlin (1974) 339-349.

[Tu67] J. Turner, Point-symmetric graphs with a prime number of points, J. Combin. Theory 3 (1967) 136-145.

[Wh75] D.E. White, Counting patterns with a given automorphism group, Proc. Amer. Math. Soc. 47 (1975) 41-44.

[Wi74] D. Wille, Note on the enumeration of self-complementary m-placed relations, Discrete Math. 10 (1974) 189-192.

[Wi78] D. Wille, Enumeration of self-complementary structures, J. Combin. Theory Ser. B 25 (1978) 143-150.

[Wo78] N.C. Wormald, Achiral plane trees, J. Graph Theory 2 (1978) 189-208.

[W76] E.M. Wright, The asymptotic enumeration of unlabelled graphs, Proc. 5th British Comb. Conf. 1975 (C. St. J.A. Nash-Williams and J. Sheehan, eds.) Utilitas, Winnipeg (1976) 665-677.

[W-A] E.M. Wright, Burnside's lemma: An historical note, J. Combin. Theory Ser. B, to appear.

OVALS IN A PROJECTIVE PLANE OF ORDER 10

John G. Thompson

Department of Pure Mathematics and Mathematical Statistics,
University of Cambridge.

In this note, I would like to record some calculations I have made
in studying the following question:

Does there exist a set S of 99 fixed point free involutions on 12
points such that for each involution (ab).(cd) which moves just 4 points,
there is a unique s in S which has {a,b} and {c,d} as orbits?

If there is a projective plane of order 10, and if that plane has
an oval, the answer to the above question is yes. This note is thus
perhaps no more illuminating than Bertrand Russell's discussion of the
present king of France, but this is not certain.

Let $\Omega = \{\infty,0,1,\ldots,9,X\}$ be a set of cardinal 12 and for each integer
≤ 6, let C_k be the set of involutions in Sym (Ω) which move precisely
k points. Let M_k be the permutation module for Sym (Ω) with C_k as a
$\mathbb{Z}$-basis, the group acting by conjugation. If $s = (a_1 b_1)(a_2 b_2)\ldots$
$a_6 b_6) \in C_6$, let

$$\lambda(s) = \sum_{1 \leq i < j \leq 6} (a_i b_i)(a_j b_j) \in M_2.$$

Then λ extends to a homomorphism of M_6 into M_2 which commutes with the
action of Sym (Ω).

Let $\nu(M_2)$ be the norm element of M_2,

$$\nu(M_2) = \sum_{t \in M_2} t.$$

For each subset S of C_6, let

$$\sigma(S) = \sum_{s \in S} s \ \in M_6$$

$$\phi(S) = \lambda(\sigma(S)).$$

With this notation, the initial question may be reformulated as:
Does there exist $S \leq C_6$ (of cardinal 99) such that
$$\phi(S) = \nu(M_2) ?$$

In an earlier paper [1], I showed that $\nu(M_2) \in \mathrm{Im}(\lambda)$. This
encouraged me to pursue the matter.

Convert M_2 to a positive definite lattice by letting C_2 be an ortho-
normal basis. For each $S \leq C_6$, let
$$d(S) = (\phi(S) - \nu(M_2), \ \phi(S) - \nu(M_2)),$$
and call $d(S)$ the discrepancy of S. Thus, if
$$\phi(S) = \sum_{t \in C_2} a(S,t).t,$$

then

$$d(S) = \sum_{t\varepsilon C_2} (a(S,t) - 1)^2.$$

A subset S of C_6 of cardinal 99 is called *special* if and only if $a(S,t) \varepsilon \{0,1,2\}$, $\forall t \varepsilon C_2$.

It is a feasible calculation to check that the set listed in Table 1 is special, and has discrepancy 330. I found it without the use of a computing machine, starting from an almost random special set of discrepancy 572. The initial set was invariant under conjugation by a subgroup of Sym (Ω) of order 11. This made it relatively easy to find. I'll describe the method I used to go from 572 to 330 with the hope that someone will make use of a computing machine to decrease further the discrepancy.

Let S be the set of all special subsets of C_6. Convert S to a directed graph by writing $S \rightarrow S_1$ if and only if the following two conditions hold:

(a) $S \cap S_1$ has cardinal 98.

(b) $d(S_1) \leq d(S)$.

Given S, the problem is to find all S_1 with $S \rightarrow S_1$. The obvious approach is to pick an $s \varepsilon S$, delete it from S, and adjoin to S an arbitrary $s_1 \varepsilon C_6 \backslash S$. The number of such ordered pairs (s,s_1) is $99(|C_6|-99)$, and $|C_6| = 10395$. For each such choice, one could compute the discrepancy of $S \backslash \{s\} \cup \{s_1\}$, and list the solutions to (b). There is a better way.

If $s \varepsilon C_6$ and $t = (ab)(cd) \varepsilon C_2$, say that s contains t if $\{a,b\}$ and $\{c,d\}$ are orbits of s; and say that t is contained in S.

Given a special set S, and an s in S, define the weight $w_S(s)$ to be the number of t in C_2 which are contained in s and in precisely one other element of S. That is,

$$w_S(s) = \text{card}\{t \varepsilon C_2 | a(S,t) = 2, s \text{ contains } t\}.$$

Let

$$o(S) = \{t \varepsilon C_2 | a(S,t) = 0\}$$

be the set of elements of C_2 which are contained in no element of S.

I also need the immediate and intermediate neighborhoods of an s in C_6. The immediate neighborhood of s is the set of s' in C_6 such that ss' εC_2. The immediate neighborhood of s has cardinal 30. The intermediate neighborhood of s is the set of s' in C_6 such that ss' has cycle types $3^2 1^6$. It has cardinal 160.

Lemma. *Suppose* $S \rightarrow S_1$. *Let* $D = S \cap S_1$, $S = D \cup \{s\}$, $S_1 = D \cup \{s_1\}$. *Then one of the following holds:*

(1) s_1 *is in the immediate neighborhood of* s.

(2) $w_S(s) \geq 5$ *and* s_1 *is in the intermediate neighborhood of* s.

(3) s_1 *contains at least* 8 *elements of* $o(S)$.

(4) $w_S(s) \geq 7$.

The proof can be safely omitted. It has an elementary combinatorial flavor.

In studying $S \to S_1$, we take the 4 possibilities in turn. Each s in S has 30 immediate neighbors, so we need to test 30.99 (=2970) cases. If $s \epsilon S$, $s_1 \epsilon C_6 \backslash S$, let $\alpha_S(s,s_1)$ denote the number of t in C_2 such that t is contained in both s and s_1 and in addition $a(S,t) = 2$. Given s_1 in the immediate neighborhood of s, we have $S \to S_1$ if and only if the following conditions hold:

(a) If $t \epsilon C_2$ is contained in s_1 and $a(S,t) = 2$, then t is contained in s.

(b) If $o_S(s_1)$ is the number of elements of $o(S)$ which are contained in s_1, then $w_S(s) + o_S(s_1) \geq 9 + \alpha_S(s,s_1)$.

More precisely,

$$d(S) - d(S_1) = 2(w_S(s) + o_S(s_1) - 9 - \alpha_S(s,s_1)).$$

Similarly if s_1 is in the intermediate neighborhood of s, then

$$d(S) - d(S_1) = 2(w_S(s) + o_S(s_1) - 12 - \alpha_S(s,s_1)).$$

To handle (1) and (2), it was necessary to have explicitly in hand the set $o(S)$. Since $d(s) = 2.\text{card}\, o(S)$, and since at all times, I had $\text{card}\, o(S) \leq 286$, this was not unduly onerous.

In handling (3), I made use of an easy result on graphs. If Γ is a graph with 6 vertices and at least 8 edges, then Γ contains a circuit of length 4. Suppose then we have an s_1 in C_6 which contains at least 8 elements of $o(S)$. Let $s_1 = \alpha_1 \alpha_2 .. \alpha_6$, where the α_i are pairwise disjoint 2-cycles. Connect i and j if and only if $\alpha_i \alpha_j \epsilon o(S)$, obtaining a graph with 6 vertices and at least 8 edges. By the preceding remark, there is a circuit of length 4. My method was to find all the sets of four 2-cycles $\alpha, \beta, \gamma, \delta$ such that

$$\alpha\beta, \beta\gamma, \gamma\delta, \delta\alpha \, \epsilon \, o(S).$$

Given such a set, there are precisely 3 elements of C_6 which contain $\alpha\beta\gamma\delta$. In no cases I have examined were there more than 50 tuples $(\alpha, \beta, \gamma, \delta)$, so no more than 150 elements s_1 came into play.

As for the possibility that $w_S(s) \geq 7$, these unusual s must be treated one by one. In Table 1, there is precisely one such s. It has weight 7 and is s_{56}.

This, then, was my procedure. I built chains $S \to S_1 \to S_2 \to \ldots$, and in each case, I found an $n \leq 7$ such that $d(S_n) < d(S)$. I then took this s_n as a new "base point", and started afresh.

In several cases, I did not strictly follow this procedure, but interpolated" between S and S', by finding an S_1 such that

$$\text{card } S \cap S_1 = \text{card } S \cap S' = 98$$
$$d(S_1) - d(S) = d(S_1') - d(S') = 2.$$

In Table 1, I omit from each s_1 the orbit containing ∞, and I omit parentheses. Thus (∞0)(12)(34)(56)(78)(9X) is written 12.34.56.78.9X.

Table 1

A special set $S = \{s_1, \ldots, s_{99}\}$ of discrepancy 330

1.	07.12.39.46.58	34.	12.38.47.56.9X	67.	08.16.3X.45.79
2.	06.23.47.59.8X	35.	02.3X.49.58.67	68.	08.19.27.4X.56
3.	02.15.34.69.78	36.	09.15.28.3X.47	69.	09.15.2X.38.67
4.	07.1X.36.45.89	37.	06.15.24.7X.89	70.	03.1X.26.49.78
5.	02.18.39.56.7X	38.	08.17.26.35.9X	71.	02.14.37.5X.89
6.	08.13.29.4X.67	39.	0X.19.28.37.46	72.	06.13.25.48.9X
7.	08.12.34.5X.79	40.	01.2X.39.48.57	73.	0X.17.26.34.59
8.	0X.19.25.36.78	41.	03.12.4X.59.68	74.	16.28.39.47.5X
9.	03.15.27.46.9X	42.	05.14.23.6X.79	75.	07.12.3X.48.56
10.	0X.14.26.38.57	43.	07.16.25.34.8X	76.	18.23.4X.57.69
11.	01.25.37.49.68	44.	09.18.26.37.45	77.	01.23.45.78.9X
12.	04.1X.29.37.58	45.	04.25.38.6X.79	78.	05.24.39.6X.78
13.	0X.18.25.47.69	46.	15.29.37.48.6X	79.	01.24.56.79.8X
14.	01.38.46.59.7X	47.	08.14.25.39.7X	80.	05.14.28.67.9X
15.	08.12.49.57.6X	48.	0X.13.24.58.69	81.	03.16.2X.57.89
16.	07.19.23.5X.68	49.	02.13.47.68.9X	82.	06.13.28.45.7X
17.	06.18.2X.34.79	50.	04.16.2X.59.78	83.	09.17.2X.46.58
18.	03.17.29.45.8X	51.	08.1X.23.46.57	84.	08.17.24.36.59
19.	04.12.35.69.7X	52.	1X.24.35.68.79	85.	09.1X.28.34.56
20.	14.27.3X.59.68	53.	08.2X.37.45.69	86.	09.13.46.5X.78
21.	04.17.23.56.89	54.	01.29.36.47.5X	87.	06.17.38.49.5X
22.	05.13.26.47.89	55.	06.1X.27.39.58	88.	09.14.27.36.8X
23.	17.25.3X.46.89	56.	02.35.46.79.8X	89.	06.19.24.3X.57
24.	04.28.36.57.9X	57.	03.27.48.5X.69	90.	07.1X.24.38.69
25.	04.19.35.67.8X	58.	05.17.39.4X.68	91.	03.19.47.58.6X
26.	01.26.3X.58.79	59.	01.2X.35.47.68	92.	03.14.28.56.79
27.	05.12.37.69.8X	60.	02.17.36.4X.58	93.	01.27.34.6X.89
28.	09.16.23.48.7X	61.	09.16.25.37.4X	94.	05.19.2X.36.48
29.	19.26.34.58.7X	62.	02.1X.48.59.67	95.	04.15.26.39.8X
30.	02.19.38.45.6X	63.	07.13.28.59.6X	96.	0X.15.36.48.79
31.	07.13.2X.49.56	64.	0X.18.27.35.49	97.	05.16.27.38.49
32.	03.18.24.5X.67	65.	05.18.29.3X.46	98.	0X.12.39.45.67
33.	06.14.29.35.78	66.	0X.29.34.57.68	99.	07.26.35.4X.89

(The orbit containing ∞ is omitted from each s_i.)

REFERENCES

[1] Thompson, J.G., Fixed point free involutions and finite projective planes, Durham Conference on Finite Groups, ed. M. Collins, 1978.